STRIVING FOR SAFETY LLC

Don't Let It Fall

Stop Dropped Objects, Save Lives

Contents

Preface

The Plant Manager, Project Manager, Construction Manager, and I (HSE Manager) gathered to meet with the Vice President of Global Operations. Every quarter, she reviewed all serious safety incidents in her plants around the world to make sure they would never happen again. Our project rarely received an invitation to discuss injuries, but we often had a near miss or two to explain, mostly dropped objects. We joined these meetings with mixed emotions, ashamed of letting one of our workers down, but proud of how we learned from the incident.

Meredith, the VP, was prepared and jumped right to the point. "I read your summary. How is the gentleman who was nearly hit by the scaffold plank doing? I can only imagine how terrified he felt. Does he still work on the project?"

"Thanks for asking," said Jesse, the Construction Manager. "The incident shook him up, but he still works for us. In fact, he asked his manager if he could speak to the workforce at the next site-wide safety meeting. He has become a passionate safety advocate at the plant. The other workers listen to him since he is one of their own."

Meredith continued, "The near miss is regrettable, but you learned from it and shared the lessons with other groups at the plant. I will make sure our other plants implement the lessons learned. I notice your absence from these meetings for the last two quarters after attending every one last year. How have you reduced serious dropped objects so dramatically? We need to share the secret."

Sandra, the Project Manager, responded this time. "It's no secret, just old-fashioned hard work. We have reduced serious dropped objects by over 80 percent. We purchased a lot of materials, such as tool lanyards, tool bags, and tarps, and retrained everyone. But the key is that our inspection staff, team leaders, and safety inspectors visit the front-line every day to check if the dropped object controls are in place and effective. If not, they find out why and fix it. Everyone on the team contributes, but Arnold drives the entire program."

Meredith brought the discussion to a close. "Well, all of you should be proud. Arnold, thank you for keeping our workers safe every day. Your work could not be more important. You may have saved a life or two over the last year. You will receive some calls from your peers at the other plants. Please share your hard-earned lessons so they can improve even faster than you did."

"Thanks, Meredith," I replied with a strange mix of pride, modesty, and regret. "I wish we had started earlier. I will be glad to help the other plants. I don't want them to learn the hard way as we did, or worse yet, hurt someone along the way."

One of the most challenging aspects of safety leadership is that you never know what would have happened without such an intervention. However, considering our previous rash of dropped objects, we may have saved a life. Meredith had praised me with the ultimate compliment for a safety professional. She reminded me why I chose the profession, why I am never satisfied, why I was not always the most popular teammate, why I answer the phone at 2 AM on Sunday morning, and why I was not on the career fast track. It was all worth it! All the leaders and teammates who contributed to this improvement should be just as proud, even prouder than of reducing costs and delivering ahead of schedule. Keeping people safe is the most difficult challenge leaders face.

Our team labored relentlessly to achieve this dramatic improvement, and we enjoyed even more success in the following year. I helped others with similar challenges and learned about the unique difficulties they faced. The purpose of this book is to help you and your organization reduce dropped objects, prevent injuries, and avoid some of the sleepless nights and trial and error we endured.

***

For updates on new resources and tools that support this book and upcoming books and events, please sign up for my email list here (https://strivingforsafety.mailerpage.com/).

Acknowledgement

Thanks to all of my former colleagues and business partners for sharing and listening.

Thanks to those working at the front-line, producing the goods and services we all need while protecting yourselves from an endless array of hazards.

1

Introduction

I'm not one that needs photos of blood or broken bones to motivate me to prevent people from being hurt. I read enough incident reports to know human beings will make mistakes and the consequences can be grave–too many cases where a few feet here or a few seconds there can make the difference between going home alive or not. Our safety controls can fail. Multiple layers of protection are needed so that no one is hurt when mistakes are made or safety controls fail. In other words, fail safely.

In addition, the mitigation of the risk of dropped objects is less understood than control of other hazards, such as confined space entry and working at height. I worked with leaders who understood the risk but not the controls, and others who denied it was even a problem.

These challenges motivated me to write this book, because lives are at stake!

I responded to many close calls with dropped objects under my watch, causing me to redouble my efforts. Some of these incidents affected me more than others: the ones on my project, the ones involving my teammates, the ones I saw. It shouldn't be that way, but it was.

Ted, a safety inspector on my team, worked at a site that was expanding a tank farm as part of a larger project. He married his bride just before the project began. His job was to observe work practices at the front-line, coach workers to work safely, and identify site-wide improvement opportunities.

He departed the construction office every morning with a full backpack, knowing he should carry everything he needed until lunchtime: his snacks, water bottle, notebook, extra gloves and safety glasses, and treats for workers deserving more than a pat on the back. One morning, he was inside of a nearly built tank approaching a work crew dismantling a scaffold. He did not see the mechanic and his helper preparing to remove a hatch on the roof of the tank.

Ted heard, "Watch Out!", and then a big *CLANG* right behind him. He turned to find a large wrench ten feet away and a corresponding dent in the steel tank floor.

Two scaffolders saw him cower and ran over to him. "Are you OK?" They relaxed as they saw him straighten up.

"Yeah, yeah. I wasn't hit. I just ducked when I heard the warning. I'm not sure why. It wouldn't have done any good. Did one of you drop that wrench?"

"Oh no! We tie our wrenches off with lanyards. I think it came from up there," both pointing to the hole in the roof.

Two wide-eyed faces stared at them through a hatch on the roof of the tank, one covering his mouth with his hand. "Are you alright?"

one of them yelled, his voice echoing in the tank. A friendly gesture, but a little too late. They had no idea Ted and the scaffolders were below them.

Ted and his wife were expecting their first child in six months. As he looked back and forth between the wrench and the hatch, he wondered how his yet-to-be-born son or daughter would fare without a father. Ted worked for me. The site manager and I were responsible for making sure he was around to witness the birth of his child. This incident, and a few others like it, led me and the team to change how we managed the risk of dropped objects at that site and others in the project. To make a long story short, it worked, but more on that later. Ted experienced the birth of his son six months later, as he should.

This book presents the fundamentals of preventing harm from dropped objects. It includes a menu of controls and practices that can significantly reduce dropped objects at your site. Specifically, the book covers the following aspects of dropped objects:

- The extensive scope of the dropped object problem.
- The potential risk of dropped objects.
- The different types of dropped objects.
- Controls to prevent the different types of dropped objects.
- How to apply a Safety Management System to bolster these controls.
- Where else to go for help.
- What people in various roles can do to prevent harm from dropped objects.

The FIVE STEPS to reduce harm from dropped objects are:

1. *Identify your priority **focus** area.* There are three types of dropped objects (from elevated work, elevated equipment, and material movement).

Focus on the most important first.

2. *Document and **communicate** mandatory requirements.* How do workers know what is expected of them if you don't tell them?
3. ***Resource** the plan:* Provide materials and people to enable implementation at the front-line.
4. ***Lead** through coaching and verification:* Initiate the change and actively support it.
5. ***Learn** and adjust:* Monitor the effectiveness of the change and adjust as needed.

The primary target audiences for this book are leaders accountable for ensuring their workers go home safe every day and the safety managers and staff that help the line managers achieve this lofty goal by providing subject matter expertise, administering the support systems, and coaching. *Leaders* are numerous in large organizations. This book will be most helpful to those three to four layers above the front-line workers. Job titles vary between organizations, but examples include maintenance managers or supervisors, operations managers or supervisors, construction managers or superintendents, project leads or managers, site managers for small to medium sites, and regional managers responsible for a number of smaller sites. People in these roles not only need to understand the big picture but also need to know enough detail to provide direction on implementation and verify the effectiveness of controls. Leaders of very large sites, projects, or organizations will also benefit from the book. They typically do not need to know some of the details presented here, but need to understand the complexity and breadth of the problem so they can decide on priorities, demonstrate commitment to improvement, and provide resources. Little progress occurs at the lower layers without that.

* * *

I have heard many excuses that stymied progress at sites. After reading this book, you should be able to challenge the skeptics in your organization who offer similar excuses.

- "That could never happen! Nothing could fall from here."
- "The area was barricaded. No one could have been hurt."
- "My inspectors will find any potential dropped objects while performing their structural inspections. Why change anything?"
- "But I am not working at height. I am on a permanent platform."
- "Our equipment is not so high. Dropped objects is not a problem here."
- "All of our people use tool lanyards. All they have to do is go to the tool room and ask for one."

"How is this book different from others on the subject?" Well, I have yet to find a single, comprehensive book on dropped object prevention. A variety of blog posts, recommended practices, incident reports, and promotional materials are available from vendors and consultants. Also, industry organizations, such as the Dropped Objects Prevention Scheme (DROPS), the Global Offshore Wind Health and Safety Organisation (G+), and the International Association of Oil and Gas Producers (IOGP) provide references, guides, and networking opportunities. Some government agencies have regulations on the hazard with corresponding learning materials. Links to some useful resources are provided at the end of the book. Most of these resources provide general summaries, detailed information on specific types of dropped objects, or guidance on industry-specific issues.

Dropped objects is just one of hundreds of safety hazards leaders must manage. Most safety books aim to help leaders improve safety culture or overall safety performance. Those are important and necessary, but this book focuses on how to eliminate harm from dropped objects. Whatever your safety leadership approach, your organization still needs the detailed technical knowledge of the hazards they face to protect themselves from harm. This book provides much of that for dropped objects. That being said, many of the concepts presented here will help you reduce other safety incidents as well, concepts such as focus, leading indicators, risk-based decision making, learning from incidents, and contractor engagement.

The multiple dimensions of dropped objects and the dozens of good practices may initially be overwhelming. Don't worry, no one uses them all, especially at first. Eliminating dropped objects is hard work, and you will have to make changes to be successful. Steven Covey once said, "To achieve goals you've never achieved before, you need to start doing things you've never done before." Start the journey by committing to continuous improvement and implementing the most beneficial practices or controls. This book will help you select the focused actions that will improve performance the most. In other words, it can help you save lives!

* * *

Notes:

- Let me clarify some terminology before we go further. I generally use the term *dropped objects* in this book. Others use the term *falling objects*. Consider them interchangeable for the purposes of this book. *Falling objects* is a broader term. All dropped objects fall, but not all falling objects are dropped. A light fixture with the corroded nuts and bolts isn't dropped; it just fell. *Dropped objects* is more common, so I will generally stick to that.
- The word, *you,* has two different meanings depending on the context. In most cases, it denotes a person leading an improvement effort at a facility or in a business. In other cases, it refers to front-line workers or supervisors, typically when describing specific controls for dropped objects.

Personal reflection and learning

Many people learn more by reflecting and practicing versus reading. At the end of each chapter, I have included reflection questions and activities to deepen your understanding of the material and to help you begin preparing your improvement strategy. I encourage you to give them a try.

- Why are you reading this book?
- Have you heard or said any of the excuses listed above? If you heard them, how did you respond? If you said them, or even nodded your head as you read them, do you truly believe them? Why? What messages might you be giving to others in your organization?

2

Why Is It So Hard?

I could have titled this book *5 Easy Steps to Eliminate Dropped Objects in 30 Days*, but I did not want to mislead potential readers. There is no quick fix. Eliminating dropped objects is hard work, but it will all be worthwhile. You have taken the first big step by committing to learning more about dropped object prevention. You have recognized a problem and are determined to improve, two key ingredients to driving change.

Dropped objects injure about 45,000 workers in private industry in the United States every year.[1] About 250 of those are killed. Unfortunately, the numbers of such tragedies have not changed much over the past five years.[2] Why do these incidents still occur so frequently? With all the managers, safety professionals, equipment vendors, industry associations, and government agencies working on the problem, shouldn't the injuries be declining?

The bad news is that it is a difficult problem. In fact, dropped object prevention is composed of several difficult problems. The good news is that these problems can be solved. Leaders, engineers, safety professionals, front-line workers, and suppliers have developed specialized equipment and work practices to address multiple aspects of the problem. By implementing these proven solutions effectively, you can reduce the harm from dropped objects at your site. Others have done it, and so can you.

Why is it so hard?

Safety is always hard

Hazards abound. Humans make mistakes. No one can live in a bubble, and even in bubbles, hazards are present. You are in business to make a living for your family, make money for the owners or shareholders, and contribute to society, not solely to prevent safety incidents. Workers are faced with conflicting priorities every day.

The consequences of failure in safety can be irreversible. If the company misses a sales target this quarter, it can make up the difference next quarter. Lose a life, it's gone forever. Reducing dropped objects will be much easier if you are already an outstanding safety leader.

It's not just one problem

Dropped objects fit into three categories, those from elevated work, elevated equipment, and material movement. Different controls mitigate the risk of each type. Stop scaffold materials from falling, and then light fixtures start to fall. Or forklifts begin to knock crates off of the warehouse shelves. Or inspectors begin losing their hardhats. All three types of dropped objects have controls, so the entire problem can be solved over time. But tackling the entire problem at once is difficult and may lead to frustration and failure. You may need to stagger your efforts, focusing first on the type of dropped object posing the most risk at your site.

There is no rulebook yet

Surprisingly, most safety-related regulatory agencies have few specific requirements governing dropped objects. As a result, few companies specifically address the controls in their procedures and training. In many countries, dropped objects are among the top five leading mechanisms of occupational fatalities.

The Occupational Safety and Health Agency (OHSA) in the United States has a few simple requirements regarding dropped objects in their walking and working surfaces regulations (29 CFR 1910.28 and 1910.29). Similar rules exist for scaffolding and other applications. See the language in the Code of Federal Regulations for the specific and current requirements. Pages of specific requirements exist for other serious hazards, such as confined space entry, electricity, falls from height, and excavations. Yet dropped objects injure more people than some of those hazards.

Many organizations are familiar with some of the controls needed to prevent harm from dropped objects. Several recommended practices are available for free (see the Resources section at the end of this book). However, with so many mandatory safety requirements for other hazards, *recommended* can morph into *optional*. Leaders and safety departments may not want to add another procedure to their bulging safety manual. They may try to manage the risk through safety moments, signs, or procedures focused on other hazards. Some workers will get the point and apply the controls. Others may think, "It's not written in the procedure, so I won't get in trouble if I don't do that. There are already too many safety rules for me to do my job."

The risk is not intuitive

Everyone should recognize that a twenty-foot piece of twelve-inch diameter pipe falling from ten feet could kill them. But what about a six-inch stud bolt from 150 feet? Or a scaffold clamp from fifty feet? The answers would range widely. All are dangerous, but can everyone be expected to know that without training? It's not intuitive, partly because few people have had the misfortune of experiencing such a variety of real incidents. Fortunately, a simple dropped object risk calculator is available to help workers assess the risk and plan their work.

Everyone has to get it

Dropped objects can affect nearly everyone. Anyone working at elevation, even just walking around with a hard hat and radio, can create dropped objects. And anyone below can be hit. Even those working in the office can be hit by falling ceiling panels and boxes stored on the top of bookcases. They all need to understand the risk and the controls needed for common types of dropped objects.

Dropped object prevention is like hand safety in this respect. Everyone can hurt their hands and needs to know the different types of hazards to hands and how to protect themselves. Let's contrast this with confined space entry. Few people (if any) are exposed to confined spaces at a site on any given day. Specific controls are required, but only the exposed workers and their supporting cast need to understand the details.

Everything can fall

Essentially everything at height can fall, and it probably has somewhere. Making a complete list for most sites is impossible, but knowing the most common dropped objects at your site and in your industry helps focus efforts where the risk is highest.

Gravity never sleeps

A gasoline leak at ambient conditions with no source of ignition will not create a fire or explosion. If an object becomes free at height, it will come down. Unlike sources of ignition, gravity is always present. The only question is whether an unfortunate person will be hit and injured or killed.

* * *

OK, it's hard. Other parts of your job are hard. Even though you know better, perhaps you had a sliver of hope for a silver bullet for dropped object prevention when you bought this book. Well, there is no silver bullet, but other people in your same predicament have been successful in reducing dropped objects. You can be successful too. Hamilton Holt said, "Nothing worthwhile comes easily. Work, continuous work and hard work, is the only way to accomplish results that last."

Summary

- Preventing dropped objects is hard; otherwise, they would have stopped a long time ago.
- Different types of dropped objects have unique controls, dropped object scenarios are endless, and everyone in the organization must be committed to and involved in the solution.
- Others have been successful; you can too.

Personal reflection and learning

- What is the largest obstacle to reducing dropped objects at your site? Why? What tactics have you already tried to overcome that obstacle?
- What other tough problems have you conquered? How did you achieve success? How can that tactic be applied to dropped object prevention?

3

How Bad Could It Be?

I had no idea!

Hundreds of fitters, welders, electricians, and painters sat in straight lines on a cool, brisk morning at the construction site.

A few of them whispered, "The site manager must be delivering his quarterly pep talk today."

"Yeah, I guess we're going to be here a while."

A surprise was in store this morning. They started most mornings with toolbox talks in groups of six to ten near their work location for the day. But in front of them was an unusual sight, a PVC tube rising from a plexiglass box near the ground to the top of a scaffold tower one hundred feet tall. The plexiglass box contained a small watermelon wearing a hard hat.

A member of the safety department at the top of the tower held up a six-inch stud bolt that a safety inspector had found during the night shift on a scaffold seventy-five feet above the ground. He dropped it in the tube.

Seconds later, the hard hat banged against the inside of the plexiglass box as the watermelon sitting underneath cracked open and sprayed its red juice all over the plexiglass. Well over half of the workers uttered different versions of *Ooh* or *Ah,* mixed with a few expletives. Some closed their eyes. A few jumped up for a better view. The site manager standing twenty feet to the right of the box flinched, anticipating the juice splattering on his face. But he never said a word; he didn't need to. The message was crystal clear.

As the workers filed out, they passed the table next to the box where the hard hat now rested, cracked in four different directions. The cracks in the watermelon were wider and still oozing sweet, red juice. Next to the cracked hard hat lay an assortment of other materials found the evening before during a *hazard hunt* for potential dropped objects: a screwdriver, a short scaffold pole, and a coiled extension cord.

A pipe fitter leaned over to his helper and whispered, "Wow, I had no idea something so small could do so much damage."

"Now I see why they keep telling us to use those tool bags, tarps, and tool lanyards. Let's stop by the tool room to get a couple of those new bags with zippers," his helper replied.

The fitter nodded. "Yeah, we will be using dozens of stud bolts just like the one the safety guy dropped in the chute. I wonder if we left that one behind yesterday?"

"I hope not. I couldn't live with myself if someone was hurt by something like that. Let's double check before we leave this evening."

Mission accomplished!

This demonstration plays out at sites around the world every week. You can find a good example made for video here (http://www.youtube.com/watch?v=6wQKUDX7D94&t=2s). If you want to hear the *Oohs* and *Ahs*, I'm sure you can find others with a quick online search. It's not the same as being there, but may be the most feasible option for small worksites.

These demonstrations and other risk awareness tools are essential for building respect for the risk posed by dropped objects. But how do you estimate the risk without cracking hard hats and watermelons?

The Dropped Object Risk Calculator

Fortunately, an industry organization called DROPS (Dropped Object Prevention Scheme, http://www.dropsonline.org/), has developed a simple calculator to estimate the risk posed by dropped objects. DROPS was founded by the oil and gas drilling industry in 1998. While they initially focused on the drilling segment of the industry, they have since widened their scope to the entire industry and beyond. They provide several useful tools for anyone trying to reduce dropped objects. The materials are free and can be adapted to the user's needs. This won't be the last reference to DROPS resources in this book.

One appeal of the calculator is its simplicity. The weight of the object and the height from which it fell (or could fall) are the only inputs. No calculus required. Not even the calculator on your smartphone. Could it get any simpler?

The calculator is available in several easy-to-use formats, including colorful charts in Imperial and Metric units and in a spreadsheet. The charts can be used to create posters in the office, a large display board at the worksite, or pocket cards for workers. They can also be incorporated into a pre-job risk assessment form.

The instructions list the assumptions behind the calculator and other useful tips. Use the calculator as a guideline, considering any exceptional risk factors that may exist. In most cases, use the result as is. The calculator assumes a hard hat is worn. The instructions urge you not to subtract the height of a person from the fall height since the calculator already accounts for that. It also assumes the dropped object is blunt, so if a sharp object, such as a pair of scissors, has fallen, consider bumping up the risk level from what the calculator shows. If scissors nick the wrong blood vessel, the outcome could be tragic. Use common sense instead of trying to find loopholes. Don't allow games to be played with the classification; they just distract from managing the risk.

Let's test your intuition with a few examples. Read the scenarios and guess what potential level of injury (fatality, days away from work case, medical treatment case, or first aid) could result. Then check the DROPS calculator chart below to see how good your intuition was. My answers are shown after the chart.

1. A 10-pound light fixture falls 12 feet from the exterior of a building.
2. A 12-ounce brick falls 50 feet.
3. A 4-ounce screwdriver falls 120 feet.
4. A 300-pound piece of sheet metal falls 1 foot to the ground while being lifted.
5. A 15-pound blind flange falls 10 feet.

* * *

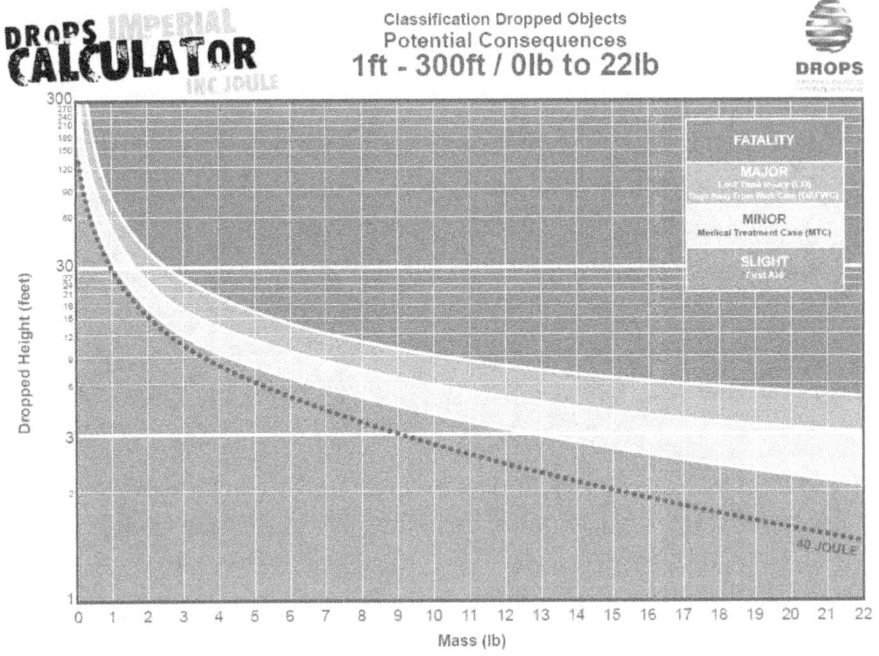

Source: Dropped Object Prevention Scheme [3]
(https://www.dropsonline.org/resources-and-guidance/drops-calculator/)

For those reading this in black and white or who are color blind (like me), the colors and categories are, from top to bottom: Red/Fatality, Orange/Days Away From Work Case, Yellow/Medical Treatment Case, Green/First Aid.

Note the steep slope of the curves at the left of the chart and how they converge. Once above about one hundred feet, nearly any object will create a risk of fatal injury. Gravity is a powerful force. An object falling from one hundred feet will be traveling fifty five miles per hour when it hits the ground (not accounting for air resistance), similar to a car traveling down the highway. Keep this in mind when pursuing opportunities to decrease work at height. Reducing work at extreme elevations should be a high priority.

* * *

And the answers are:

1. A 10-pound light fixture falls 12 feet from the exterior of a building: *Red or Potential Fatality.*
2. A 12-ounce brick falls 50 feet: *Yellow or Medical Treatment Case.*
3. A 4-ounce screwdriver falls 120 feet: *Red or Potential Fatality.* The chart says Yellow or Medical Treatment Case. However, a screwdriver is not a blunt object, and the outcome could be much worse if it struck a person point down. I would estimate the risk as Red or Potential Fatality.
4. A 300-pound piece of sheet metal falls 1 foot to the ground while being lifted: *Orange or Days Away From Work Case.* This is a tricky one since it requires some judgment. It is off the chart, but if you use the spreadsheet version, the result is Red or Potential Fatality. In my opinion, the worst credible outcome is the sheet landing on someone's foot or ankle. Someone would have to slither underneath an active load on a crane to create a worse outcome. The lesson here is to use some judgment, but not too much. Playing games with risk rankings is a sure way to lose credibility as a genuine safety leader.
5. A 15-pound blind flange falls 10 feet: *Red or Potential Fatality.* This is another tricky example and one where judgment can get you in trouble. Did you subtract off the five to six feet of a person's normal height? If you did, you come up with Yellow or Medical Treatment Case. If you follow the instructions, the result is Red or Potential Fatality. In organizations that make a big deal of potentially fatal incidents (as they should), you may be tempted to reduce the risk ranking. Don't put your credibility on the line; just focus on learning from the incident so it doesn't happen again.

More risk, more controls

Why use a tool such as the DROPS risk calculator? Workers are more likely to put controls in place to prevent harm from dropped objects if they understand the risk. They can also look at their surroundings and visualize how badly **they** could be injured by the actions (or lack thereof) of nearby work crews. Workers will be armed with the knowledge to protect themselves and their colleagues.

Chapter 5 describes the multitude of controls available to reduce the dropped objects from elevated work. How will workers decide which controls to use and when? They deserve to know what is expected of them. A site procedure and training program should describe the expectations, but also provide the flexibility to adapt the controls to specific tasks.

The higher the risk, the more controls they should use. If a control is not feasible for their task, they should be thinking, "What else can I do to make up for it?", not "Oh well! I tried, but it doesn't work in this situation." For example, large studs and bolts can be stored in closed containers when not being used. But at some point, the workers will remove the studs and bolts for installation and won't be able to secure them. What can be done to make up for the increased risk? They may be able to cover the holes in their work surface to prevent a dropped bolt from falling below. Perhaps they can make the exclusion zone below more effective. We will explore this concept in much more detail in Chapter 5.

Pre-job dropped object risk assessment

Many people use the dropped object risk calculator for estimating the potential risk of a dropped object which has already occurred, i.e. what if it actually hit someone. Typically, this determines the classification of the incident and influences the type and depth of the investigation. A more proactive and impactful use of the calculator is to assess the risk of upcoming work so effective controls can be put in place.

> Imagine you are in a morning toolbox talk. The supervisor hands out laminated pocket cards with the risk calculator chart on one side and questions on the other.
>
> What could fall?
> How far could it fall?
> What is the risk level?
> How will you secure the materials?
> How will you exclude people from the area below?
>
> The work crew looks at their tool bags and the basket of materials ready for the crane to lift. You hear, "Well, the biggest thing I have is this wrench, and it weighs about a pound."
>
> Then a colleague says, "Yeah, but there's a blind flange we are going to remove. It must weigh over thirty pounds."
>
> "Wow, if we plot that on the chart, it is clearly in the red zone of the calculator. That could kill someone."
>
> The supervisor says, "So, what are we going to do about it?"
>
> Another crew member says, "Let's set up a good barricade down here before we go up. We can also bring a sling to tie the flange off before we remove the final bolts."
>
> His helper says, "Yeah, and let's bring some tarps in case we drop any of the studs and nuts. We already have lanyards for our tools."

Finally, the supervisor says, "Great job, team! Go ahead and set up. I'll come by before you start to verify all the controls we discussed are in place."

What a great toolbox talk! If only they were all that good. The supervisor asked a couple of open-ended questions, recognized his crew, and committed to follow up on their work.

By contrast, many supervisors will read through highlights of the Job Safety Analysis. They get to the row on Dropped Objects (if it is even included), and say, "Let's not drop anything today. Make sure you have tool tethers before going to the worksite." Next.

A sample form to guide these discussions or individual risk assessments can be obtained via the download link in the Author's Note. The form does not address all aspects of the hazard but is a good starting point. It can be customized to the site or incorporated into the site's existing job planning processes. Changing the questions over time keeps the discussions fresh and emphasizes areas needing improvement. In any case, maintain the basics:

- What could fall?
- What is the risk?
- How will I prevent the fall?
- If it falls, how will I protect people around me?

This may sound like overkill. "If we do this for every hazard, the crews will never get to work." True, but you can do so for one or two hazards where significant improvement is needed. This risk assessment tool is particularly effective when beginning an improvement program. That's why focus is so important. Once a dropped object prevention culture is established, this planning will be a habit and not need a form. By then, you may need a form for another focus area.

Summary

- The risk from some dropped objects, especially smaller ones at great heights, is not intuitive.
- Fortunately, we have a very simple and effective tool to assess the risk of potential dropped objects before work begins and the potential risk of dropped objects which actually occur.
- The level of risk is instrumental in planning the controls needed to protect people from harm.
- More risk, more controls.

Personal reflection and learning

- How have the risk assessment examples above changed your understanding of dropped object risk at your site?
- How do your work crews currently assess the risk of dropped objects from their work?
- Apply the risk calculator to past dropped objects at your site. In hindsight, how does the level of investigation compare to the potential risk of the incident?
- Visit the worksite and complete a pre-job dropped object risk assessment with a work crew. How did the crew react? What did you learn?

4

Three Types of Dropped Objects

One of my safety inspectors, Freddie, stepped into my office shaking his head and flicking his finger across his smartphone screen. "You won't believe what just happened!" he said.

"Was anybody hurt?"

"No, no, I didn't mean to scare you. It's just that we were doing so well on stopping dropped objects and this happens."

As he shoved his phone in my face, I pulled the reading glasses out of my pocket and saw a windsock and its frame lying on a slab of concrete with a fresh, nearly white divot a few inches away.

"Oh my gosh, we were lucky. Was anybody nearby?" I said, knowing windsocks were always on the top of the production modules. It must have fallen one hundred feet or more. The windsock itself weighed a few ounces, but the frame was heavy and dangerous.

"Yeah, one of our operators, Jess, was walking back to the control room. It missed him by ten feet."

"Wow, he was lucky! I'm sure the close call shook him up, but at least he gets to go home tonight. Please send me the photo. I need to give the boss a heads-up," I said, dreading the trip down the hallway.

I walked into the plant manager's office, trying not to scare him like Freddie did to me.

"Jim, we just had a high potential dropped object, but nobody was hurt." I handed him my phone so he could see for himself. "Jess, one of our operators, was ten feet away. He's a little shaken but grateful to be unhurt," I said.

"Glad to hear that. I'll ask Jess' manager to give him the rest of the afternoon off after the nurse checks him out. He needs time to settle down. Do we know why it fell?"

"Not yet, but nobody was working on the upper deck, and no lifts were in progress. It appears to have just worked loose. We are gathering the facts and statements and will begin the formal investigation in the morning."

Jim asked, "Was this on our dropped objects checklist?"

"I'm afraid not. We focus our inspections on work activities at height since that is the source of most of our dropped objects."

"Let's add it to the list."

"Let's wait until the investigation is finished, then we can revisit the follow-up actions," I said as I prepared to leave.

I knew this was a different kind of dropped object than the type we had focused on for the last year. We were making excellent progress, but we had more to do to solve the entire dropped objects problem. Our dropped objects checklist was working so well the plant manager may have thought it would eliminate all dropped objects. We would have that difficult conversation later.

Three problems in one

Dropped objects can be classified into three types, or sources:

- Elevated work
- Elevated equipment
- Material movement

There is no right or wrong way to categorize dropped objects. The primary purpose of the categories is to break a large, diverse problem into parts you can focus on and tailor solutions to. Applying one broad solution to the problem may lead to the situation with the windsock illustrated above.

The conventional categories for dropped objects are *static* and *dynamic*. The oil and gas drilling industry, one of the most advanced in the battle against dropped objects, established these categories years ago. The categories work well for them, but confuse some people outside of the industry and don't fully represent the breadth of the problem. Before going further, let's look at the definitions of the categories in each classification scheme.

The "Dropped Object Prevention Scheme Recommended Practice" published by DROPS defines *static* and *dynamic* dropped objects[4]:

Static: "Any Dropped Object whose failure may be attributed to gravitational or natural forces (i.e., without an applied force, unsecured items, or failure of fixings)."

Dynamic: "Any Dropped Object whose failure may be attributed to applied forces (e.g., from the impact of equipment, machinery or other moving items, severe weather, helicopter downdraft, manual handling, etc.)."

The categories used in this book are:

Elevated work: Dropped Objects originating from elevated activities of workers, such as scaffolding, maintenance, and operations.

Elevated equipment: Dropped Objects originating from elevated equipment with no direct involvement of workers, such as light fixtures, ceiling tiles, windsocks, and valve handles.

Material movement: Dropped Objects originating from transfer of materials from one location to another, such as from cranes, gin wheels, hand-carrying, trucks, and forklifts.

Elevated work

These dropped objects result from the **direct actions of workers** while working at an elevated location.

Think of *work* broadly here. It could be as simple as walking on the second floor of a warehouse to check inventory. Their hard hat might fall. They might kick a wrench lying on the grating to the first floor. But the most common sources of these dropped objects are scaffolding (erection, dismantling, modification), maintenance activities (lubricating equipment, installing blinds, replacing valves, removing equipment, painting), construction (installing new equipment, building structures), and inspection.

Also, think of *elevation* broadly, not only working from scaffold platforms and permanent platforms on elevated decks but also working on ground level next to an excavation or equipment pit.

Examples of such dropped objects include scaffold components, hand and power tools, nuts and bolts, piping components, personal accessories (hard hats, mobile phones), construction aids, motors, valves, masonry products, lumber, and replacement parts. The list is endless.

The controls to prevent these dropped objects are directly applied by workers before and during their task, for example, using tool lanyards, tying red tape around an exclusion zone, and placing a tarp over the grating. The principles of risk assessment and human performance are keys to success here.

The controls for preventing dropped objects from elevated work can be categorized using four layers of protection, **DROP**:

- **D**ECREASE the amount of work done at elevation as much as feasible.
- **R**ESTRAIN any loose tools, materials, and equipment being used at elevation. ,
- **O**BSTRUCT any holes, openings, or gaps in the work area.
- **P**REVENT people from being in the line of fire of a potential dropped object.

Elevated equipment

Unlike the previous type, these dropped objects occur **without direct human interaction**. They come free for the reasons listed below, and gravity does the rest. Dropped objects from overhead equipment used for moving materials (e.g. cranes) are generally included in the material movement category since the causes are different.

The risk factors leading to these types of dropped objects include:

- Corrosion, caused by dissimilar metals, harsh environments, and chemical exposure.
- Vibration, caused by rotating equipment, fluid flow, and wind.
- Improper installation, not following installation instructions or not using securing devices.
- Collision, for example, with a load being moved by a crane.

Examples include light fixtures, piping and cable supports, public address speakers, insulation materials, nuts and bolts, out of service equipment, valve handles, door components, ceiling tiles, and roof hatches. This list is also endless.

This is an **asset integrity problem**, so the controls used in those programs can be effective here. Such controls include verification that all elevated equipment are designed and installed correctly, use of Reliable Securing, focused and systematic inspections, and maintenance management. These controls all involve human tasks as well. However, the tasks are performed by many different people over the life of the equipment versus tasks performed by a small crew during the work itself, as done for elevated work.

Material movement

This type of dropped object may be the most diverse. As the title indicates, these dropped objects originate from movement of materials from one location to another. *Movement* can mean hand-carrying materials, using heavy industrial equipment such as cranes and hoists, and using vehicles such as flatbed trucks and forklifts. The controls are **mostly related to load securement and proper inspection, maintenance, and operation of heavy equipment**. Robust procedures and programs exist for these operations, such as lift plans, load securement procedures, operator training and certification programs, and rigging procedures. Unfortunately, having procedures and implementing them effectively are two different things.

Once again, the list of these dropped objects is endless. Examples include drilling tubulars, large machinery, pallets, chain hoists, scaffold components, and rigging components.

So what!

Why do we need these categories? Why not just use Static and Dynamic? The controls for each category are quite different. A *one size fits all* approach will generally be unsuccessful.

Going back to the story about the windsock, their improvement plan focused on dropped objects from elevated work. Early in the project, most of the dropped objects consisted of tools, scaffold components, and construction materials. To reduce those dropped objects, the project team focused on human behaviors, i.e. verifying the effectiveness of dropped object controls at the front-line and addressing the root causes of any shortfalls. But the windsock was a different type of dropped object, one from elevated equipment. The approach did not address equipment mounted at height. No work was being done near the windsock, and the checklist included nothing on the controls needed to prevent this incident.

Focus is needed to make significant and near-term improvement. Understanding the most common type of serious dropped objects at your site is essential so you can focus on the controls for that type first. Once you make sustainable improvement, you can do the same for the other types of dropped objects. Along the way, a dropped object prevention culture will develop, making the subsequent steps easier.

Let's go back to the nomenclature of the categories for a minute. Static mostly corresponds to dropped objects from elevated equipment, and Dynamic mostly includes dropped objects from elevated work and material movement. Dynamic is subdivided further because the controls are so different. No matter what categories are used, they will overlap. That is OK; just focus on the controls. For instance, if an anemometer falls off the boom of a crane during a lift, is it from material movement or elevated equipment? If the root cause is related to improper installation or lack of inspection of the anemometer versus the act of lifting, it fits best in the elevated equipment category. These borderline calls will be the exception rather than the rule, so don't let them become a distraction.

Summary

- Dropped Objects come in three types, from elevated work, elevated equipment, and material movement, all with different controls.
- Dropped objects from elevated work result from direct actions of workers
- Dropped objects from elevated equipment occur without direct interaction of workers. This type of dropped object is an asset integrity problem.
- Dropped objects from material movement are diverse, all the way from hand-carrying to large cranes.
- Breaking dropped objects into categories helps leaders focus improvement efforts where they matter most.

Personal reflection and learning

- Did any of the three types of dropped objects surprise you? Why?
- Obtain a list of dropped objects at your site for the previous year. Categorize them into the three types described here.
- What type of dropped object presents the most danger at your site? Why?
- How strong is your current dropped object prevention culture? What signs do you see?
- Go to the worksite at the beginning of the shift and participate in a toolbox talk. Ask the crew the following: What could fall? Where could it go? How can they control the risk? How well did the crew participate? How well did they identify the potential dropped objects?

5

Dropped Objects from Elevated Work

Think broadly

If you anticipate a potential dropped object, you can prevent it. Some controls protect against a variety of dropped objects no matter what they are, where they fall from, and how they fall. An exclusion zone is a good example. Whether a worker loses their grip on a hammer, kicks a loose toeboard, or poorly prepares a load for the crane, no one should be hurt if the area below is barricaded and no one enters. However, most controls must be applied closer to the source of the dropped object to be effective. Think **broadly** about potential dropped objects so all the necessary controls can be identified. Let's explore two mindsets, *"I'm not 'working at height'"* and *"that could never happen."*

"I'm not 'working at height'"

I met a pipe fitter, Olga, installing a reconditioned valve in the middle of a large grated platform sixty feet above the ground. Despite my red hat (indicating a dreaded *safety guy*) and her not being a native English speaker, she seemed eager to talk to me and smiled from ear to ear. Stud bolts, nuts, a blind, and a few of her tools rested on a thick tarp under the valve, but other tools and a large bolt lay on top of the grating. A three-inch wide gap surrounded a pipe support that passed through the grating just one foot from the tarp.

After we introduced ourselves, I asked, "What could fall down during your work? What might happen if it fell?"

She pointed to the tarp and said, "Nothing is going to fall. Plus, I'm not working at height. I don't need to use lanyards on my tools."

I pondered that for a moment and then pointed to the guardrail for the platform about fifty feet away. "Can we walk over there together? It will just take a few minutes. I don't want to slow you down, but this is important."

Olga hopped up and walked with me, welcoming the opportunity to stretch her legs. We looked over the guardrail to the ground sixty feet below. The plant was in the midst of an extensive maintenance campaign, so we saw a dozen people walking from module to module at ground level. Olga raised her eyebrows and covered her wide open mouth with one of her gloved hands. She got the message without me saying a word.

As we walked back to her work location, her smile returned. She was pleased to learn something new. I stopped ten feet short of her work area and peered through the grating. Olga followed my line of sight, and her smile disappeared. Another crew was underneath her work area, performing the same task as her.

She said, "Oh no! If I drop something, it could hurt one of those guys. They would never see it coming."

"So, what will you do now?"
She paused for a moment. "One of us needs to stop work and go somewhere else. Let me find my supervisor so he can reassign one of us. You know what? You safety guys are not so bad after all." It was my turn to smile!

In her mind, she was not *working at height*, at least when I first arrived. Guardrails surrounded the platform, so she did not need to wear fall protection. Therefore, she didn't need to control dropped objects either. She was new to the site, so maybe that was the message she took away from the rushed and information-packed safety induction a few days earlier.

People can be injured by dropped objects anytime they are underneath any activity at height. The work may be on a permanent platform or a green-tagged (safe to use) scaffold platform. It might even be on ground level next to an excavation or equipment pit or next to a flatbed truck being unloaded. Dropped objects can injure or kill in any of those situations.

In addition, *work* doesn't have to be work as you might think, with wrenches, bricks, and grinders. Talking to people, inspecting equipment, and reading a gauge are enough *work* to lead to dropped objects: hard hats, radios, notebooks, cameras, flashlights, loose items on grating, etc.

I use the term *elevated work* versus *work at height* deliberately. Think broadly about *working at height*!

"That could never happen!"

Two mechanics, Carlos and Tommy, were preparing to reinstall an access panel on a machine they had repaired. The steel panel was one-eighth of an inch thick, twelve inches by twelve inches, and weighed just over forty pounds.

A safety inspector, Marie, approached them as they were about to start. "How are you guys doing this morning?"

"OK, I guess. We're almost done. We just need to install this panel before handing the equipment back over to Operations. They were checking on us all day yesterday," Carlos replied.

"Let me guess! You want to see our Dropped Objects Risk Assessment? Let's make it quick," Tommy said as he reached into the front pocket of his coveralls and handed the form to Marie.

"Thanks. We can take time for safety, you know, especially your safety," Marie said as she began reading the risk assessment. "I see you identified your wrench and some nuts and bolts as potential dropped objects. Thanks for using lanyards on your wrench and keeping those nuts and bolts in your backpack. What else might fall?"

"That's all we could think of or we would have put it on the form," Tommy replied. "Can we get back to work now?"

"What about that panel? It's not secured to anything while you put it back in place?"

Carlos replied this time, "Come on! Sure, we could drop it, but where is it going to go? It's way too big to fit through any of the holes in the grating."

"I agree, but you are working next to a guardrail. Couldn't it go through there?"

"You're really stretching it now, Marie. *There's no way that could happen.* What's it going to do, bounce off the grating four feet to the left? I don't think so," Tommy said as he took two steps to the guardrail and looked down. "Plus, see that scaffold platform along the deck below; it would stop the panel in the one in a million chance it falls over the edge."

Marie saw the ground was forty feet below and noticed two electricians organizing their materials in the distance. "Wouldn't it be easy and safer to install some netting along the guardrail and barricade the area underneath?"

"Like I said, Operations needs this done yesterday. Didn't you see the barricade below? The electricians put it up yesterday. Are we done now? Tommy, give me a hand."

Marie took a couple of steps back, shaking her head and taking some notes in her tally book.

As Carlos tried to position the panel, he caught his finger between the panel and the machine, flinched in pain, and let go of it. Tommy tried to catch the panel, but deflected it through the adjacent guardrail instead. Marie screamed–then ground her teeth.

They all leaned over the guardrail and watched in disbelief as the panel slid through a one-inch gap between two scaffold planks on the first deck of the tower, and in horror, as it did the same on the next deck. Before the panel even hit the ground, Carlos and Tommy each thought, "This can't be happening to us."

The two electricians had moved closer to the module but were unaware of the work above them. They heard someone yell, "Watch out!" Before they could even look up, they heard and felt a "thump" as the panel hit the ground. They turned and saw the plate fifteen feet away in a cloud of dust. It had fallen within their barricade where they thought they were safe. Instead, they were just lucky.

With their knees still shaking, the mechanics started down the scaffold stair tower to apologize as the electricians ran up the stairs to give the mechanics a piece of their mind. Cooler heads prevailed as they met on the second platform. Tommy and Carlos then looked up to Marie who was thinking, "Told you so," but didn't dare say so.

Still not convinced?

A rigger was connecting a trash container to the crane hook along the edge of a platform one hundred feet above ground. A stair platform was ten feet below him. He lost control of a shackle as he was trying to insert the pin. The shackle bounced off of the handrail next to him, hit the outer handrail of the platform below, and fell the remaining ninety feet to the ground. What are the odds? No one was hurt in that incident either. Barricade tape surrounded the area, but people commonly took shortcuts through the area. The number of shortcuts plummeted–at least for a while–after supervisors shared the incident in the toolbox talks the next morning.

If you anticipate what might fall and where it might go, you can prevent it from hurting someone. If not, you are relying on luck, which is not an effective risk control. Don't experience the same dangerous learning curve others have. Someone could lose their life in the meantime. The lesson here is to think "It **could** happen to me… how can I stop it," instead of "that could **never** happen."

How could the mechanics have prevented the panel from falling so close to the electricians in the first example? Give it some thought and check yourself when we revisit the scenario later in this chapter.

Layers of protection - DROP

People will make mistakes, so our systems must fail safely where possible. Preventing harm from dropped objects associated with elevated work relies on worker behaviors, i.e. attaching tool lanyards, covering grating, and avoiding the line of fire, all of which can fail. Also, the controls protect other workers more than the one putting the controls in place, perhaps making the risk less personal. The persons underneath the work are the ones injured.

Since no controls, other than elimination of the hazard, are perfect, multiple layers of protection are needed to fail safely. Controls can be divided into four layers of protection with the mnemonic, **DROP**.

- **DECREASE** the amount of work done at elevation as much as feasible.
- **RESTRAIN** any loose tools, materials, and equipment used at elevation.
- **OBSTRUCT** the falls of any objects beyond the immediate work area.
- **PREVENT** people from being in the line of fire of a potential dropped object.

Controls higher on the list are generally more effective. The further down the list you go, the less control you have on the outcome, and more people are at risk. In simpler language:

- **D**ECREASE:If it doesn't go up, it can't fall down.
- **R**ESTRAIN:Don't let it fall in the first place.
- **O**BSTRUCT:If it starts to fall, don't let it fall far.
- **P**REVENT:If it falls, make sure no one is hit.

Why four layers? Workers may forget to apply one or two layers or not apply them effectively. Some layers may not be feasible for the assigned task and location. Applying two, three, or even four layers of protection increases the likelihood of failing safely. They also provide a menu of controls for workers to select from in proportion to the risk.

Let's explore each layer in more detail.

DECREASE

Elimination tops the hierarchy of controls for hazard management. While not all work at elevation can be eliminated, every little bit removed reduces the risk of dropped objects and the risk of people falling from height. For each task eliminated, the other controls described below are not necessary. This often saves money as well. The planning needed to achieve these reductions may occur years in advance for large scopes of work or minutes before the job for smaller tasks. Several examples are shown below, but also consider how you can modify your existing work processes to accomplish the same objective.

<u>Modify designs and activities to reduce work and equipment at elevation</u>

A HAZID, or HAZard IDentification, is a brainstorming exercise used to identify safety risks associated with an activity, project, or piece of equipment. Once the hazard is identified, the existing controls and potential additional controls are identified as well. It is usually a broad discussion aimed at inventorying all hazards and controls, but can be repurposed to identify opportunities to reduce exposure for one hazard, such as work at elevation.

These reductions may be realized during construction, operations, and maintenance.

Start by gathering a diverse group from Engineering, Construction, Operations, Maintenance, and Safety. Discuss the construction process, operational procedures, and expected maintenance tasks for each piece of equipment in a section of the plant being built or modified. Ask questions such as:

- What can fall?
- How can the design be modified to eliminate the risk?
- How can the construction methods be modified to reduce work at elevation?
- What maintenance or inspection tasks are required after installation? How can they be done at a lower elevation? Can any be eliminated?
- How will the equipment and materials be moved to the worksite? What safer alternatives are feasible?
- How should accessories be fastened so they do not become loose and fall?

Think outside of the box and capture all ideas. There are no stupid questions or suggestions in these exercises. Even if an idea clearly won't work, the suggestion may spur better ideas from others. All items will be evaluated in more detail after the workshop. Resist the temptation to say, "we haven't done that before" or "that will never work", thereby stifling creativity. In order to reduce exposure, you will need to make changes.

Other hazards can be addressed at the same time to add more value. Just add a few more questions to the list of prompts. Try to limit yourself to three hazards or fewer so you don't lose focus.

After the workshop, a smaller group evaluates the pros and cons of each idea. They will use criteria such as the amount of risk reduction, impacts on other safety hazards, life-cycle costs, equipment functionality and reliability, and constructability. Closely scrutinize activities that are frequent or performed at extreme heights since they have the most potential for risk reduction. Many ideas will be discarded quickly, but some should lead to worthwhile changes.

Examples of outcomes of a HAZID include completing installation of smaller components on modules or skids before lifting into place and selecting equipment types that can be inspected and maintained from lower elevations. A specific example of the latter type is changing the light bulbs on the light poles along highways by lowering them to the ground versus climbing or using boom trucks. These choices can also reduce life-cycle cost by eliminating the cost of accessing the elevated location.

Plan simultaneous operations (SIMOPS)

Simultaneous operations (or SIMOPS) are activities occurring at the same time that may affect each other in some way. In the context of dropped objects, one crew working above another crew is a simultaneous operation. If a person above drops something, a person below could be injured. When people are working over others, they cannot use an exclusion zone below the work, one of the key controls for preventing harm. Lifting loads over active work areas is another example.

> Celia stood outside of the manway of a large storage tank being inspected. She was the confined space attendant, keeping track of the traffic in and out of the tank, logging gas test results, and on standby for emergency calls. She saw Stuart, the safety inspector, cresting the stairs over the tank dike. He came by twice a day to see how they were doing. She was new to the site, so she appreciated the extra pair of eyes.

Instead of walking toward her calmly as he normally did, he waved his arms and yelled at someone on top of the tank. Celia stepped six feet further from the tank and looked up. Directly above her, two workers with puzzled looks were gesturing back at Stuart. Stuart pointed to the curved stairway on the side of the tank, and they finally started moving toward it.

Stuart walked up to Celia, and she asked, "Hi Stuart, what's all the commotion about?"

"When I climbed to the top of the dike, I noticed those two welders directly above you. It looked like they were setting up to repair a guardrail. If they dropped something, it could hit you on the head."

Celia winced. "Oh no. Thanks for catching that. The blowers being used for ventilation are loud, so I didn't hear anyone up there. My first priority is taking care of the inspectors inside the tank, and I didn't notice anyone climbing the stairs."

The welders walked up to Celia and Stuart briskly. "What is it now? We've been waiting on parts for the handrail for a week. We were just getting ready to work when you waved us down."

"I understand your frustration. I just noticed that you were setting up to work directly over Celia. What would happen if you dropped something?" asked Stuart.

"We have tethers for our tools, but we can move ninety degrees around the tank," replied one of the welders.

"OK, please do so, and also, please set up a barricade under your work. I realize there are not many people around, but others might not notice you up there like Celia didn't." Stuart continued, "I didn't see your supervisor at the SIMOPS meeting yesterday afternoon. If he had been there, he would have known about this potential conflict and reassigned you to another job. I'll give him a little coaching later today."

Two hours later, Celia heard a big *thump* off to her left. She walked out twenty feet from the tank and looked toward the barricade the welders had set up. She gasped when she saw a two-foot piece of two-inch pipe sticking out of the ground. She looked up at the welders who both shrugged their shoulders, and one of them yelled, "Sorry!"

Once Celia's heartbeat subsided, all she could think about was how she would thank Stuart. If he had not come by, that pipe might have hit her instead of the dirt. Also, since he asked them to set up a barricade, no one else was near the falling tube.

Many sites use a structured process for reviewing planned work for SIMOPS, often as part of a *Permit to Work* system. These processes can identify and prevent situations where work plans include people working over others. SIMOPS planning does not reduce the amount of work at elevation, but reduces the risk to those below by excluding them from the line of fire. The first check can be done when the plans are being created, but another check the day before the job is worthwhile in case plans change. Work crews should check one last time just before starting work. The further ahead the planning is, the more time you have to rearrange work and minimize impacts on cost and efficiency. Due to the late intervention in the story above, the welders would be losing several hours of productive time.

Laminated drawings of buildings, modules, or units (plan views and elevation views) are useful in facilitating the discussion. Each work group marks their work locations on the drawings. Supervisors and planners identify potential conflicts and discuss and resolve them on the spot. Digital tools are also available.

Bring only necessary materials to elevated work locations

Poor housekeeping is a common cause of dropped objects. When workers take only what they need to their elevated work location, with reasonable contingencies, materials will be easier to store during the work and remove afterwards.

This can be a fine balance. For example, a stockpile of frequently used materials, such as scaffold components, at an elevated location can reduce the overall risk versus lifting them up and down every time they are needed. Such storage areas should be tidy, well-organized, and protected from objects falling through holes or guardrails.

Preassemble on the ground

Preassembling equipment or structural components on the ground before lifting reduces the number of tools, nuts/bolts, and smaller components that could fall while doing the assembly at height. This includes reassembly of equipment that was broken down for shipping, when feasible. In some cases, preassembly can increase the risk of a dropped object during lifting or other movement, for example, a light pole sticking far out from a module. Therefore, the risks of movement and preassembly need to be assessed in order to determine the safest option.

A project to fabricate an industrial building benefited from such an exercise. The conventional construction method was to fabricate small sections near the ground and lift them into place. By the end, the lift would be over 120 feet. The project team decided to fabricate larger sections, requiring less work at higher elevations to weld the sections together and connect the utility lines. This site had struggled to prevent dropped objects early in the project. As you know now, most things dropped from 120 feet are dangerous. No one will ever know, but this decision may have prevented a serious injury at the rate things were falling earlier in the project.

Check tools and materials before going to elevated work locations

I once observed a mechanical technician, Tony, checking nuts and bolts over a small tarp near the top of a stairwell ninety feet above the ground. I introduced myself and asked, "What is your primary task for today?"

"I am installing brackets on the column over there."

"What are you doing right now?"

"Well, the area is hard to access, and I will be wearing fall protection. I'm just making sure I have everything I need and that all the threads are OK," he said with a smile and then continued screwing nuts onto bolts. Perhaps he thought his response would satisfy me, and I would move on.

I was impressed and thanked him for his forethought, but was not finished with him. "Oh, sounds like a great idea. It will be safer for you and anyone who may wander underneath. I see you are using a tarp to prevent the nuts and bolts from falling through the grating. That's a good practice. What else can you do to prevent the nuts and bolts from falling?"

"I don't know. I am being careful here."

I offered a suggestion to him. "Would it be possible to check them on the ground next time, so you eliminate the risk of dropping one? Can you do that for your colleagues?"

Tony shrugged his shoulders and said, "Sure, that would be easier and safer. I don't want to hurt anybody."

I finished with more recognition. "Thank you for planning ahead. I also see lanyards on your tools. That's fantastic. Have a safe day."

Other examples include testing power tools, checking battery charge, counting spare discs and blades, and inspecting tool lanyards and fall protection equipment.

Attach tool tethers before going to elevated work locations

Workers occasionally drop tools while attaching tethers. They got the message and were trying to prevent dropped objects, but something fell anyway. Unfortunately, the details are important. Connecting lanyards to tools before going to height eliminates one opportunity for them to fall. Tools without built-in attachment points may need several connectors and require more effort than simply attaching a carabiner.

RESTRAIN

Tethers and containers

When thinking of dropped object prevention, tool tethers immediately come to mind. When using tools at height, they can be restrained with a tether (or equivalent) connected to the tool on one end and an anchor point on the other. Depending upon the weight and type of tool, the anchor point may be a wrist strap (for light tools), a tool belt, a structural beam, or even a tool bag. It sounds simple, but on numerous occasions, I have seen tethers connected to the tool but not an anchor point, thereby defeating the purpose.

Many tools need accessories to secure a tether. Tools with built-in holes are easily secured. Others, such as wrenches or needle-nose pliers, are more difficult. Fortunately, many vendors provide a wide variety of load-rated and tested tethers and accessories (rings, specialty tapes/wraps, velcro holsters, etc.).

Proper tethering of tools is an acquired skill. A group of people should be trained on tethering methods for the specific tools and tethering supplies used at the site. They could be tool room attendants or super-users sprinkled among different craft groups and contractors. In some cases, the supplier of your tethering equipment may be able to visit your site and demonstrate proper use of their products. A poorly secured tool may be just as dangerous as one not secured at all. As with most controls, front-line workers should inspect the tethers and attachment points for wear or damage before use.

Engineered and tested tethering products provide more protection than strings, ropes, and other homemade devices. Knots vary in quality and strength and introduce another opportunity for human error.

If feasible, fit each tool with a lanyard. Switching lanyards from one tool to another provides another opportunity for the tool to fall. Connect lanyards over tool bags, containers, or tarps if switching them between tools is necessary. Some tool bags include attachment points inside the bag so that tools can be tethered while still attached to the bag.

When tools and smaller materials are not being used and not secured with tethers, store them in a container, preferably a closed one. Purpose-built tool bags with this feature can be purchased. Many are durable, load-rated, closable, and include rings for attaching tethers.

If using other containers, shorter ones with wide bases are less likely to tip over. These can also be placed under active work to catch small objects slipping from a worker's hands. Taller containers, such as five gallon paint buckets, can become top-heavy when full and tip over when bumped. Plastic bags tear on grating and other rough work surfaces. Cardboard shipping boxes deteriorate quickly in the rain and sun.

At-height tool kits

These kits are common on drilling rigs, offshore vessels, and in some smaller facilities. The tools are ready for easy attachment to a lanyard and stored in a dedicated tool box. Many contain custom foam cutouts for each tool so missing ones are obvious. The box is typically locked and under the control of a designated person. When tools are used for elevated work, they are removed from the tool kit, logged on a register, and returned at the end of the shift. If the custodian notices a missing tool at the end of the shift, they initiate a hunt for it. Sites or vessels take these extreme measures when dropped object risk is high and scope is manageable. For large construction sites or during extensive maintenance campaigns, they may not be practical. Good housekeeping and personal accountability are even more essential in those cases.

It's not just the tools

Tools get most of the attention, but they don't always present the largest risk. Tools are typically being used on something much bigger: a pump, a valve, a structural support, a building component, a blind flange, or a light fixture. Remember the access panel discussed earlier–it weighed over forty pounds–the wrench weighed only one pound. Purpose-built lanyards and accessories for securing tools are plentiful, but few are specifically built for other materials or equipment.

Tool lanyards may work for some materials and equipment as is. In other cases, lifting appliances, such as slings and shackles, may be a solution for securing larger equipment. If installing equipment that will be lifted by a crane, you may be able to use the attached rigging until the equipment is partly installed and secured in place. If removing something to be lifted later, you may be able to attach the rigging before removing all the fasteners and secure to a nearby anchor. When cutting materials, any pieces that could fall should be retrained before the cut is made. Most of these materials and equipment are bigger and heavier than tools, so secure them to something strong enough to support the maximum expected load, i.e. not a person.

Unfortunately, some materials cannot be tethered at all times, e.g., nuts, studs, bolts, and gaskets. They should be stored in a closed container while not being used. At some point, they must be taken out of the container to be installed and often cannot be tethered. In these cases, the *Obstruct* and/or *Prevent* controls must be used. Scaffold components are another common example. They can be secured in racks, boxes, or bags when being moved to the work location, but they are not typically secured during final assembly. These examples further illustrate why so many layers of protection are needed.

Personal accessories

Some sites take extraordinary steps to tether even the smallest tools, but then allow people to work with mobile phones in open pockets or unsecured hard hats. Since these are *personal accessories*, it sometimes gets *personal* when asking people to secure them. It's amazing how passionate people are about using a chinstrap on a hard hat. Other accessories include flashlights, gas detectors, tablets, and tape measures. Some are not heavy or a serious risk at low heights, but they are prevalent and can be dangerous at greater heights. The exposure adds up.

Vendors supply a plethora of solutions for these: lanyards and chin straps for hard hats, closeable holsters for radios, and tethering kits for many others. A simple closed pocket will work for small and light accessories.

The curve on the dropped objects risk calculator becomes steep around thirty feet. When people are accessing heights above this, it makes sense to secure these accessories. Instead of taking the protection on and off, why not get in the habit of securing personal accessories at all times? It sets a good example for others and reduces the risk of connecting and disconnecting the securing devices. It's hard to walk up to an electrician on the third deck and give him a hard time about not restraining his screwdriver when your hard hat is unrestrained and your radio is in an open shirt pocket.

OBSTRUCT

This layer of protection covers all holes, gaps, and other pathways to areas below the immediate work area. If an object starts to fall, it is stopped immediately so that no one below can be injured.

"But you just told me I need to restrain everything. Nothing is going to fall. Why must I do more?"

Unfortunately, it is not so simple. Lanyards aren't always attached securely, bags are left open, buckets tip over, and loose items get kicked down the stairs. In addition, recall those nuts, bolts, and scaffold poles that are not restrained at all times. This control stops falling objects before they build dangerous momentum. The dropped object remains in an area the worker controls instead of relying on others to respect a barricade below. However, don't let the presence of this control lead to complacency in restraining things. Heavy or sharp objects can still injure people in the immediate area.

A former colleague told me about an incident near a water storage tank. A pipe fitter was removing a thirty-six-inch diameter manway cover. Most large manways have davits, or swinging metal arms above the manway, to support the weight of the manway as it is moved away from the tank. This reduces the risk of manual handling and eliminates the need for a crane. In this case, the bolt attaching the manway to the davit was severely corroded, and the davit fell on the person's foot, leading to permanent damage.

Don't underestimate the risk of short falls of very heavy objects.

<u>Holes, gaps, and penetrations in work surfaces</u>

Cover these pathways unless it is impossible for anything dangerous to fall through. If you hear "that could never happen," feel free to tell one of the stories in this book, or better yet, tell your own. Many grated surfaces are full of holes by design. Scaffold planks don't always fit together snuggly, especially around curved structures or equipment. Piping and electrical cables pass through floors or grating, leaving a gap around the perimeter. Dropped objects will find these gaps; or it sure seems so.

Tarps, fine mesh netting, rubber mats, and plywood are commonly used to cover such holes. It sounds simple, but this control is difficult to implement well. Many of these gaps have odd shapes that defy a standard solution. In addition, tarps and mats are hard to secure to equipment such as a large pipe passing through grating. If not secured well, a heavy object will pass right through a tarp as if it wasn't even there. Objects may become lost in the folds of a tarp or net, so take care when moving them. Custom wood cutouts can be effective for odd shapes, but require more effort to install and can become dropped objects themselves.

I once saw a bright yellow tarp that was stretched taut across some open grating as two technicians worked over it.

"Hi! My name is Arnold. I am visiting to review your dropped objects program," I said, as I offered my gloved fist for a bump.

Matt and Alexa took me up on the offer and introduced themselves.

"What objects could fall during your work today?"

Matt jumped right in and said, "We are replacing the pressure gauge right here," placing the palm of his hand on top of the gauge. "So this gauge could fall once we disconnect it, and the new one could as well. We could also drop our wrenches. We have a few spare fittings in the bag on the tarp."

"I see you have lanyards on your tools. Thanks for doing that. I presume this yellow tarp is here to stop those gauges from falling, or your tools if they get loose. How do you keep it so taut?" I said.

Alexa pointed to a velcro strap attached to the tarp and wrapped through a hole in the grating.

"Now that's a great idea! How did that come about?"

Matt and Alexa looked at each other with slight grins. Matt finally broke the awkward silence, "Well, if she's too modest to say it, I will. Alexa came up with the idea. She made one in the shop, tried it out a few times, and showed it to our supervisor. He gave her some budget to make more. Our shop is pretty busy, so she arranged for a vendor down the road to make them for us."

"Alexa, show Arnold how else we can use this."

Alexa removed another tarp from its dedicated shoulder bag, walked over to the guardrail, strapped one end to the top rail, and the other to the grating. "If we were working near the guardrail, we could do this to prevent things from falling over the side."

"Wow. That's pretty awesome. I hope you received some recognition. Do you mind if I take a photo so I can show people at other sites?"

"Go right ahead. We're proud of Alexa and the design. And yes, our supervisor took the whole team out to lunch to recognize Alexa. She's going to have his job someday!"

Alexa turned around as she tried to hide a giggle with her hand.

"I'm so glad I stopped by. I learned something new today that will help others. Have a safe and productive day."

Covering a work surface may create slip, trip, or fall hazards. The risk can be reduced by making the covers visible and securing them against movement. For holes large enough to step through, use materials rated for the expected load. Before covering any work surface, make sure it has no defects, such as corroded steel or rotten wood. Once covered, any problems with the underlying surface will be out of sight and out of mind for nearby workers and passers-by.

For scaffolding, make sure the tolerance for gaps is low. Recall the story about the panel falling through two levels of scaffold. If the gap was smaller or covered, the panel would not have fallen to the ground next to the two electricians. Inevitably, odd-shaped structures will cause gaps that should be covered.

Toeboards

Toeboards on permanent platforms and scaffolds are required in most areas. They prevent materials from being knocked off the edge of the platform. Workers should verify they are present and secure. If loose, they can become quite dangerous dropped objects themselves. If they are not present, take time to decide whether to postpone the work or install a temporary barrier around the work area.

Guardrails

Dropped objects don't always fall straight down. When work occurs near a guardrail at the edge of a platform, many sites secure fine mesh netting, fencing, or purpose-built barriers from the grating or toe-board to the handrail. This prevents objects from being flung through the rails or bouncing off the floor and over the toe board. Recall the story about the falling access panel. A co-worker deflected it through the rail when he tried to catch it. He did not intend for the panel to fall over the edge.

Ensure the type of material used to cover guardrails does not catch enough wind to cause loading above design specifications. Otherwise, the structure could be damaged, thereby creating other hazards. This is why netting and other *open* materials are often used. Temporary barriers require maintenance, especially when in place for weeks or months. The wind can tear the netting and other workers will remove sections to do their jobs. In either case, will the unsecured barrier stop a flying hammer?

More permanent coverings may be in order for areas with a lot of work or high foot traffic. If handling unrestrained materials next to and above a guardrail, consider extending the barrier above the top rail with scaffold poles or equivalent.

Minimize storage of materials near guardrails as this leads to a lot of material handling in a high-risk location. These areas are tempting, since they are easier to access with cranes and often away from primary pathways. If such a storage area is necessary, place a sturdy barrier underneath and along the sides.

For guardrails along busy paths, consider adding a barrier to stop items lost from open pockets and snagged tool bags from falling to the next level or further.

Boom lifts, scissor lifts, and other mobile elevated work platforms have guardrails around their work baskets. Some manufacturers and third parties sell kits to line the basket to prevent objects from falling through the rails. Consider asking the supplier for this feature. Follow manufacturer's recommendations, as some solutions may create other hazards, such as excessive wind loading on the boom.

PREVENT

Prevent is the last layer of protection. "You've got to be kidding. I restrained my tools and materials, and I obstructed all pathways from my work area. How could anything get through?" First, it **can**. Second, it **will**. Some objects cannot be *Restrained*, and some areas are hard, if not impossible, to *Obstruct*. Both are true of scaffold erection.

I may not have convinced some of you yet, so let me share another incident.

> The Maintenance manager was doing his morning rounds and saw a pipe fitter preparing to install a pressure gauge on a pipeline.
>
> "I don't think I have met you yet. My name is Julio, the Maintenance manager here. It is so nice to have you on board. I'm glad to see you working on that pressure gauge. Operations has been asking about it for days."
>
> "Thanks for stopping by. I am happy to be here. My name is Gerard," the fitter replied.
>
> "Gerard, could you show me how you will prevent dropped objects

today? You are working pretty high up," Julio said as he looked over the guardrail to the ground sixty feet below. Several workers were passing through on the way to their worksites.

Gerard showed Julio the lanyard connecting his wrench to a nearby pipe support. He tapped the plywood covering the grating with his foot and pointed to the tarp covering the gap where a pipe passed through the grating.

"Wow, that's great. You must have received your dropped objects training already. Are you going to install a barricade below?"

"No, they told me in training that if I used lanyards and tarps, I didn't need a barricade."

"Oh. That's right. Have a safe and productive day," Julio said as he continued on to the next work crew.

Unfortunately, the lanyard was loosely secured to the wrench, and the tarp was not attached to the pipe. Gerard lost his grip on the wrench. It slipped out of the lanyard and passed right through the tarp. The wrench fell sixty feet to the ground, crashing five feet from a terrified passer-by.

Gerard recognized potential dropped objects and used two independent controls. He concluded that a barricade below his work area was unnecessary. In fact, he did as he was trained. Julio reviewed his controls and supported his decision. Yet neither of the controls worked when needed. Details matter. Quality matters.

Some of you may be thinking the opposite, "I'll just barricade underneath the work, then I don't need to worry about all those lanyards, tarps, nets, and so forth. If things fall, they won't hit anybody." People who spend little time at the front-line tend to believe exclusion zones are more effective than they really are. Exclusion zones are administrative controls, i.e. procedural controls relying on human behavior, so they can fail.

Some organizations track the effectiveness of their exclusion zones when objects fall. An exclusion zone is *effective* if it was in place for the work and no one was inside. I have seen different sources of information indicating the range of effectiveness of exclusion zones at 30-70 percent. The range is wide, but how does it make you feel? Confident? Scared? Skeptical? The entire range is too high for me, considering lives and livelihoods are at stake. The numbers show that exclusion zones need to be better, and need to be combined with other controls.

Exclusion zones and barricades

Along with tool lanyards, exclusion zones are the most visible controls for preventing harm from dropped objects. Notice the wording here–they do not prevent dropped objects, only the harm the objects may inflict.

Exclusion zones come in many sizes, shapes, and colors, but the most common is red *Danger* tape hung on traffic cones, posts, or structural supports. Other examples include plastic chains, crowd control barriers, painted areas on the work surface, and scaffold rails. The oil and gas drilling industry typically uses the terms *red zone* and *no-go* zone, or the broader term, *zone management*. Exclusion zones under permanent equipment with high dropped object risk may be marked on the work surface itself.

In most cases, persons who need to enter an exclusion zone must obtain explicit permission from the designated owner of the zone. The owner should allow entry only when the hazard is not present; for example, the work is paused or the work crew moves to another location.

Why are exclusion zones less effective than many people assume? It seems like a straightforward rule. Red tape means stay out. Some of the reasons for this are cultural, the same as why people do not wear safety glasses or clean up their worksite when finishing a job. Others are more specific to dropped objects. Some people don't understand the risk of dropped objects or feel they won't be hurt by one. So many barricades may be present that it is difficult to get anywhere without crossing one. If left up for days with no apparent activity, the exclusion zones lose credibility, and people enter anyway. Sometimes, they are unlabeled or too small.

During several site visits, I walked up to the inside of a barricade wondering, "How did that happen?" My knees would become weak for a few minutes. Someone let me down and put my life at risk, and I didn't pay enough attention. It was just a matter of time before I was nearly the victim of a dropped object.

> I was nearing the end of my last site walk of a week-long dropped objects audit at a module fabrication yard. They were building components of a much larger plant that would be assembled at a site halfway around the world.
>
> "It's getting kind of late. Have you seen enough?" said my escort, Chen.
>
> "Sure. Let's head back to the construction office."
>
> Just before exiting the module, I heard *ting, tang, clang.* I knew that sound. I looked up and spun around. Before I could locate the falling object, I saw a pair of pliers hit the ground twenty feet in front of me, but just ten feet from the work crew walking ahead of us. We all looked up to locate the source of the pliers. The module was very congested, so it took a couple of minutes to find the electrician in the pipe rack fifty feet above us.

The crew ahead of us began yelling at the electrician in the local language. I did not understand the words, but I got the gist of the message.

I spun around again, more slowly this time, looking for the barricade tape. Did I just walk into another open-ended barricade? Surely there is an exclusion zone, since nothing but pipes, beams, and open air were between us and the electrician. There was no barricade at all! Chen grimaced as he noticed my disappointment. We walked over to the pliers. A four-foot piece of string was tied to a hole in one of the handles. I wasn't hit, but it was pure luck.

Chen shook his head wildly and said, "Go back to the office. I need to talk to the electrician."

In the audit closeout meeting the next day, I described my findings and recommendations using my typical structured approach. Some of the leaders appeared to be disappointed, but eager to improve. Others were quiet, shaking their heads slightly or peeking at their smartphones. I concluded with the story I described above, ending with a simple statement, "I don't feel safe walking around this site any more."

The quiet ones quickly perked up. That one statement had more impact than the previous thirty minutes of findings and recommendations. It was personal, and they knew it. The previously quiet leaders began sharing their embarrassment and sought input from those who had been engaged the entire time. I was a bystander. The leadership team now owned the problem and was determined to improve.

At a minimum, exclusion zones should be:

- Visible (right color, right height)
- Complete on all sides
- Owned by one, identified accountable person
- Labeled on all sides with the hazard, date valid, owner name, and contact information
- Large enough to account for deflections and horizontal momentum
- Removed when no longer needed

These are the minimum. In some scenarios, such as scaffold erection, this is the only control available. Since we know exclusion zones aren't perfect, consider other means to make them more effective.

So what else can be done?

- Use hard materials instead of plastic tape, for example, scaffold poles or crowd control barriers.
- Assign a dedicated barricade attendant to make sure no unauthorized persons enter.
- Identify a single, marked entry/exit point.
- Establish a standard for minimum size.
- Discourage shared sides with other work parties.
- Treat unauthorized entries as safety incidents, which are investigated.
- Keep an entry/exit log, as done for confined spaces.
- Use technologies such as digital exclusion zones combined with personal locators and alarms.

You can think of other enhancements if you try hard enough.

If this degree of rigor seems unnecessary, go outside and scrutinize your own exclusion zones. Or assign someone to survey the exclusion zones every day and count the number of unauthorized entries and ineffective or missing barricades. You might be surprised and scared.

I kept a mental tally on my site visits. At one of our plants with numerous upgrade projects and a good overall safety culture, I witnessed at least one case of work with serious dropped object risk without an effective barricade on eight out of ten visits. The tape on one side of the barricade would be laying on the ground. Unauthorized personnel would walk inside to grab something from their tool cabinet. Or there was no barricade at all. I would share these findings with project leadership before leaving the site, but was usually told that I focused only on negative observations, and it wasn't really that bad. On the last two visits, I found none. Being the *bad cop* on the previous eight visits had paid off.

I have addressed unauthorized persons in exclusion zones, but authorized persons can put themselves at risk as well. Typically, members of a work crew who review and sign off on the Job Safety Analysis or Work Permit describing a barricade as a control are authorized to enter, since they understand the hazards. For example, let's consider scaffolders.

I encountered a couple of scaffolders dismantling a scaffold during an audit of a maintenance turnaround. They used a small construction elevator to transfer materials from the fourth story to the ground, a much better alternative to handling every piece using a human daisy chain. One scaffolder on the fourth-level removed components and placed them in a rack on the elevator. The other was located on the ground and unloaded the rack. I noticed that the scaffolder on the ground stood a few feet from the bottom of the elevator while his partner above filled the rack. A visible barricade surrounded the entire area, extending fifty feet from the elevator. I motioned for the scaffolder on the ground to come to the edge of the barricade where I stood.

After introducing myself, I said, "Can we talk about your safety for a few minutes or am I interrupting your work?"

"It's OK. I need to unload the rack when it comes down, but I can talk now."

"I am pleased to see you using an elevator here. It sure beats handling all of those materials by hand, and a lot safer too. And I see a large barricade to keep other people out–such as myself."

"Thanks, it's a great tool: faster, cheaper, and safer," he said, happy to show off their proactive job planning.

"What are you responsible for while your partner is filling up the rack?"

"I'm waiting for him to indicate the rack is full, and then I lower it to the ground." He pointed to the control panel near the base of the elevator. "I unload materials from the rack to those containers over there and send the rack back up." He was confident he had all the right answers.

"Could you stand out here while the rack is being loaded?"

"I guess so. It's just a habit and saves a little walking back and forth."

"Which is safer?" I said, hinting at the right conclusion.

"Well, now that I think about it, if he misses the rack, one of those poles or planks could hit me. That would not be pretty. I don't need to be over there. I'll stand out here from now on. Thanks for the suggestion. I'll tell my partner when he comes down for a break."

> I felt proud that my coaching seemed to be effective, but had an inkling that the helper was just appeasing the safety guy and not embracing the message. My doubt was erased the next day when I encountered him again, standing outside the barricade while his partner loaded the rack. He hadn't just done so to make me happy, but to keep himself out of harm's way. He got the message. I waved and gave him a *thumbs up*. He returned the gestures, together with a smile from ear to ear! A good example of why I loved my job.

Finally, let's go back to one of the minimum attributes of exclusion zones, *large enough to account for deflections, horizontal momentum, etc.* "How large is *large enough*?" First, consider what factors may cause some objects to deflect further than others: wind speed and direction, shape and density of the object, and the number of obstacles along the fall path. A thin sheet of plywood falling on a windy day will usually be deflected much further than a twelve-inch blind falling over the edge of a module. That twelve-inch blind will usually be deflected more falling down the interior of a module filled with equipment than over the edge of an overhanging scaffold. Second, Energy Safety Canada has developed a draft tool[5] that provides more quantitative guidance for idealized scenarios. For example, the tool predicts a deflection of one hundred feet for an object dropped from one hundred feet and striking an obstacle halfway down–in other words, a two hundred-foot diameter exclusion zone. An exclusion zone of that size will be impractical in many situations, but the tool provides a range to consider. In practice, the work crew will need to consider the amount of deflection risk, the risk level of potential dropped objects, the other controls being used, impacts on surrounding work, unintended consequences of a large exclusion zone, and site layout, to implement a feasible, but protective exclusion zone.

Nets and overhead protection

Exclusion zones have been covered thoroughly because they are one of the most common, visible, and essential controls. However, they may not be feasible in some situations. For instance, work may be required over an internal road needed for emergency response or delivering critical materials. There may be work or heavy foot traffic over an access point to a building or module.

Other options are available. One is to use a horizontal net under the work area to catch things before they fall far, like the *Obstruct* controls, but further below the work area. Earlier, I talked about tarps or netting covering the top of work surfaces. These nets are located under the work area. This option may also be useful when covering gaps from the top is not feasible due to their size, shape, location, or number.

Another option is to install *hard tops* or covered walkways, often using scaffold materials with planks forming the roof. These are sometimes used on city streets where construction is done above sidewalks. If constructed well, these can provide safe passage to an access point in some cases.

These alternatives should not take the place of exclusion zones across the board, since they can fail as well. If the weave of the net is too large, small but heavy objects may fall through. If left up for a long time, the environment may cause deterioration. *Hard tops* are usually limited in size, so they may not protect people from deflection. In addition, objects may pass through *hard top* roof materials, a reminder of the energy some dropped objects possess.

What do I use when?

This chapter has covered a lot of options for controls. Having options is good, but with options come choices. You may feel overwhelmed. "How many do I use? What do I use when? What if I can't prevent access below the work?"

Many procedures and training packages do not answer these questions. They describe the controls and leave the rest to the supervisors and front-line workers. But what message does this send? Use at least one? Use all of them every time? It's OK not to use them if they won't work? Are they mandatory or optional? The front-line needs more clarity. Otherwise, work crews will choose different controls in similar situations, and some will put themselves at more risk.

A couple of options for providing this clarity are to:

1. Specify which controls are required in all situations.
2. Provide risk-based principles for selection of controls.

"What's wrong with asking people to use all the controls every time? The more the better, right?" Ideally, yes, but…first, it is not always feasible. Some objects cannot be *restrained,* and it isn't always possible to *obstruct* the path of a falling object. Second, barricades may not be feasible in some cases. Third, some small dropped objects present very little risk; three robust controls may be viewed as overkill and can have unintended consequences. By insisting that workers use all controls every time, leaders may lose credibility when the front-line workers realize they can't effectively implement some of the controls or when the dropped object risk is very low. They may conclude the leaders are disconnected from the reality of their work. They may give up when they find the first control or two are not feasible, especially if leaders walk by and say nothing.

One option is to try to predict every scenario and specify which controls to use in each. This may work for simple sites with limited scenarios, but for most sites, the procedure would be so long no one would read it—and some scenarios would still be missed.

Another option is to ask the supervisors and front-line workers to assess the risk and select the best combination of controls to minimize the potential for harm. "If I can't do A, then how can I make up for the added risk? Perhaps, I need to do B, and make C even better. Or maybe I should call time out and get help." This does not mean a *free for all*. You provide risk-based rules to ensure a proper level of protection and resources to help when needed. The front-line should know the work the best, understand the risk, and know which controls will be most effective. They and their colleagues are also the ones at risk. You should still have checks and balances to ensure the guidelines are working as intended. Some of these assurance methods will be covered in Chapter 8.

A risk-based approach uses a combination of (1) the Dropped Object Risk Calculator, (2) a job-specific, dropped object risk assessment, and (3) risk-based guidelines for selection of controls. Notice the word *risk* is in every component. The first two have already been covered, so let's learn about the last one and see how they fit together.

The risk-based guidelines might look like this:

If a dropped object from your work could cause a:

- *Fatality (Red on the DROPS Calculator): use THREE effective layers of protection.*
- *Lost Time Injury (LTI) (Orange): use at least TWO effective layers of protection.*
- *Medical Treatment Case (MTC) or First Aid (Yellow or Green): use at least ONE effective layer of protection.*

When these guidelines cannot be met, discuss and agree alternatives with your supervisor before beginning work. Use additional controls when feasible.

Or it might look like this:

If a dropped object from your work could cause a:

- *Fatality or LTI (Red or Orange on the DROPS Calculator): use as many*

effective layers of protection as possible.
- *MTC or First Aid (Yellow or Green): use at least ONE effective layer of protection, more if feasible.*

When these guidelines cannot be met, discuss and agree alternatives with your supervisor before beginning work.

These are just two examples. You may choose something different depending upon your exposure, safety culture, and local regulations. The objective is to provide clear guidance to the front-line. All controls will not be feasible for all scenarios, even some in the Red zone of the DROPS calculator. The supervisor should be consulted in such cases.

Before showing how this works in practice, let's examine a few of the terms used in the guidelines.

A *layer of protection* is a physical control, not a procedure or "I'll hold on with two hands" or "I'll find a buddy to help." *Restrain, Obstruct,* and *Prevent* are the layers of protection. A combination of tarps, boards, and netting to *Obstruct* objects from falling below the work area is **one** layer of protection. Unfortunately, *Decrease* does not count here, even though it is the most effective control. If you eliminate an exposure, it's one less time this guidance is needed.

An *effective* layer of protection (or control) will work when needed. A loose tarp over a big opening is not effective. A barricade extending only to the edges of work overhead is not an effective control (there should be an allowance for deflection). An effective tool lanyard is fit for purpose, in good condition, and secured at both ends. Quality counts as much as the number of controls. A weak control is the same as no control in many cases. The appearance of a control may appease the supervisor or the safety inspector walking by, but won't protect you and your colleagues. If the control will not work, replace it with one that will.

Finally, the phrase, *at least*, sounds very subjective. It is subjective; risk management usually is. Use as many controls as feasible and in proportion to the risk. If a dropped object could seriously injure someone, and all three layers are *feasible*, then use all three. *Feasible* doesn't mean *easy*; it means *possible*, while not introducing more serious hazards or causing significant impacts to the work.

This approach sounds complicated, but a few examples should clear it up. In this exercise, I will use the first version of the guidelines. These examples are for illustrative purposes only. The best combination of controls for each scenario ultimately depends on factors specific to the site and work location.

Scenario 1

Replacing a light fixture (fifteen pounds) near the top of a two story building (thirty feet above ground) from a boom lift.

<u>What can Fall? What is the Risk?</u>

- Light fixture - Red
- Tools (wrench, screwdriver) - Yellow
- Nuts/bolts - Green
- Personal accessories (helmet, radio) - Yellow

<u>Restrain</u>

- Light fixture - Tool lanyard, sling, or net around fixture
- Tools - Tool lanyards
- Nuts/bolts - Store in closed container while not being used; may not be feasible when in use
- Personal accessories - Closed pockets, lanyards, holsters

<u>Obstruct</u>

- May not be feasible if there is no structure between the work and the ground

Prevent

- Barricade area under work

Layers of Protection (achieved / required)

- Light fixture - 2/3
- Tools - 2/1
- Nuts/bolts - 1/1
- Personal accessories - 2/1

Comments

- They only achieved two layers of protection for the high-risk light fixture. The work team should discuss alternatives with their supervisor. Perhaps they could use a dedicated barricade attendant to bolster the effectiveness of the barricade and partly compensate for the missing control.
- They have two controls for the tools where a minimum of one is required. Since the barricade will be applied for the light fixture anyway, and tool lanyards are easy to apply, two controls are justified. That is a good example of what *at least* means.
- This is also an example of the *Decrease* layer of protection. Instead of taking hours to erect and dismantle a scaffold, a boom lift is being used for this quick and simple job.

Scenario 2

Building a scaffold in preparation to replace a light fixture from Scenario 1

What can Fall? What is the Risk?

- Scaffold components (tubes / planks / clamps / etc.) - Red
- Tools (wrench, hammer, level) - Yellow
- Personal accessories (helmet, radio) - Yellow

Restrain

- Scaffold components - May not be feasible when in use, i.e. connecting components to the rest of the structure
- Tools - Tool lanyards
- Personal accessories - Closed pockets, lanyards, holsters

Obstruct

- For the first level or two, there is no means to *Obstruct*. For higher levels, they might be able to apply netting/barriers and hole covers to the lower levels as they build. Even then, falls over the side would not be obstructed. This is not an effective control.

Prevent

- Barricade area under work.

Layers of Protection (achieved / required)

- Scaffold components - 1 / 3
- Tools - 2 / 1
- Personal accessories - 2 / 1

Comments

- The crew achieved only one effective control for the scaffold components vs. three required. Unfortunately, this is typical for scaffold erection. So, what do you do? First, remember *Decrease*. In Scenario 1, the crew used a boom lift instead of a scaffold and achieved two controls for the red risk. Boom lifts have their own serious hazards, so work planners and crews need to assess the overall risks for the alternatives at the site before selecting the means of access.

- If they decide to use scaffolding, the work crew should meet with their supervisor and discuss how to ensure no one will be in the barricade. Perhaps they should dedicate a crew member to enforce the barricade. To protect their own crew, they should pause assembly when a helper is passing materials to his/her teammates. Scaffolders on the ground should step out of the exclusion zone while assembly occurs.

Scenario 3

Removing a twelve-inch blind while working on a scaffold fifty feet above the ground.

What can Fall? What is the Risk?

- Blind - Red
- Tools (wrench, screwdriver) - Red
- Nuts/studs/gasket - Orange
- Personal accessories (helmet, radio) - Yellow

Restrain

- Blind - Before all studs are removed, secure with a shackle and/or sling, if feasible. This may be needed to lower the blind to the ground anyway. Take care not to damage the face of the blind, as that can create other risks.
- Tools (wrench, screwdriver) - Tool lanyards
- Nuts/studs/gasket - Store in closed container while not being used; may not be feasible when in use
- Personal accessories (helmet, radio) - Closed pockets, lanyards, holsters

Obstruct

- Cover gaps in or around the scaffold with boards or tarps. Secure them to the scaffold.
- Apply netting or other barrier to the sides of the scaffold.

Prevent

- Barricade the area underneath AND around the scaffold.

Layers of Protection (achieved / required)

- Blind - 3/3
- Tools (wrench, screwdriver) - 3/3
- Nuts/studs/gasket 2/2
- Personal accessories (helmet, radio) - 3/1

Comments

- Unlike for Scenario 1, the crew could apply three layers of protection for the higher risk objects since they were working on a scaffold platform vs. from a boom lift.

Scenario 4

An electrician is rewiring a breaker box on the second floor in a warehouse. She is working from open grating twelve feet above the floor. All of her materials weigh less than a pound.

What can Fall? What is the Risk?

- Tools (pliers, screwdriver, wire cutters) - Green
- Wire/connectors - Green

- Personal accessories (helmet, radio) - Green

Restrain

- Tools - Tool lanyards, may not be necessary but a good habit
- Wire / connectors - Store in closed container while not being used; may not be feasible when in use
- Personal accessories - Closed pockets, lanyards, holsters

Obstruct

- Cover the grating with a tarp. Should be easy. If not, it may not be necessary if other controls are in place.

Prevent

- Barricade if not too disruptive to others. Otherwise, it may not be necessary if other controls are in place.

Layers of Protection (achieved / required)

- Tools - 3 / 1
- Wire/connectors - 2 / 1
- Personal accessories - 3 / 1

Comments

- The crew needs at least one effective control. It is easy to get two or three for the tools, so do it.

Scenario 5

Let's go back to the panel cover that slipped through cracks of two successive decks of scaffolding (in Chapter 5). Two mechanics were reinstalling an access panel on a machine they had repaired. The steel panel was ⅛ inch thick, twelve inches by twelve inches, and weighed just over forty pounds.

What can Fall? What is the Risk?

- Panel - Red
- Tools - Orange
- Nuts/bolts - Green
- Personal accessories (helmet, radio) - Yellow

Restrain

- Panel - May not be feasible to restrain since they were in the final stages of installation. If attachment points were present on the outside of the panel, a lanyard or sling could potentially be used to secure it to a nearby structure.
- Tools - Tool lanyards
- Nuts/bolts - Store in closed container while not being used; may not be feasible when in use
- Personal accessories - Closed pockets, lanyards, holsters

Obstruct

- The panel was deflected through the guardrail as it fell. It is common practice to cover the guardrail with netting or other material for this reason. If the panel was being handled over the top rail, then the protection should be extended.
- The floor of the first scaffold platform could be covered with a secured tarp or other material.

Prevent

- The work crew should discuss the arrangement of the barricades with the electricians. If possible, they should each have their own exclusion zone.

Layers of Protection (achieved / required)

- Panel - 2 / 3 (if panel not restrained)
- Tools - 3 / 2
- Nuts/bolts - 2 / 1
- Personal accessories - 3 / 1

Comments

- The crew has only two controls for the panel vs. the required three. They may be able to attach a securing device to the panel, at least until it is partly installed. If not, the effectiveness of the barricade should be increased.

* * *

Hopefully, you now grasp how applying simple risk-based rules can lead to robust, fit-for-purpose controls for a wide variety of scenarios. In order for this to work, the workers need to understand the risk calculator and the full menu of controls. Chapter 8 will show how to prepare them.

In practice, you will probably use a combination of the prescriptive and risk-based approaches by establishing some rules for controls that should be in place the vast majority of the time. Otherwise, there may be endless, unproductive debates on common risks and controls. For instance, shall personnel secure their hard hats or not? Only while above thirty feet or all the time? Is a lanyard required or will a chinstrap do? In my opinion, this is better managed as a site-wide rule versus a person-by-person, day-by-day decision. Deviations can be approved for those rare cases where it may not be feasible. Another example is the appearance of exclusion zones. Some sites use multiple types of exclusion zones for dropped objects. A single, site-wide specification for exclusion zones of different types should reduce the likelihood of unauthorized entry.

Housekeeping

Poor housekeeping is one of the largest contributors to dropped objects. The objects can fall during the work itself or linger for weeks, months, or even years before finding their way down. All it takes is a gust of wind, a kick down the stairwell, or moving equipment with something hiding on top.

Examples of good practices for improving housekeeping are listed below. You may need to try several to prove your commitment to an orderly workplace. If one does not work, some of the others may be successful. Making the connection between housekeeping and dropped objects may give workers added motivation to embrace the need for a clean workplace. Keeping the workplace tidy is one thing, but lives are also at stake. When someone discovers a dangerous item left behind the day before, show the workforce at a safety meeting the next day. Point to the risk on the risk calculator, drop the item down the chute on a watermelon, or display it near an access point.

- Ensure containers for waste and recyclable materials are available, convenient, and emptied regularly.
- Stop work fifteen minutes before the shift ends so workers can clean up. Take away the common excuse, "You don't give me time to clean up." Housekeeping is work.
- Assign a housekeeping focal point who is accountable for a defined area (deck, module, building). They are not responsible for cleaning up, but are a focal point for troubleshooting. They should know the work parties

in the area and can encourage the supervisors to clean up their areas. If the problem is widespread, they may call a housekeeping timeout for everyone.

- Include a condition in work permits to clean the area before the permit can be closed. The issuer of the permit (or delegate) should inspect the site to confirm. Be firm about this. This approach does little good if work areas are rarely inspected and left with loose materials present.
- Conduct cleanup events, including participation by site leadership. Showcase the removed material the next day as dropped objects and injuries that were avoided.
- Encourage work teams to take only what they need to the worksite, with reasonable contingency.
- For small sites, consider using *work at height* tool kits to help identify tools left behind at the worksite.
- Recognize work teams with the best housekeeping behaviors. Buy them lunch or let them off an hour early on Friday.

Scaffolding

Scaffolding is prevalent at many sites and is a large contributor to dropped objects. During the erection and dismantling of scaffolds, materials are handled manually. Most of the components are heavy, the heights can be significant, and only one control may be feasible. A bad combination.

When you consider the amount of materials scaffolders handle, the difficult positions they work from, and the tools and equipment they carry around, they do a fantastic job. They are some of the hardest workers on a site. However, with the sheer amount of materials handled manually, components will fall. Scaffolders may appear superhuman at times, but they make mistakes like the rest of us.

Let's begin with the first and most effective layer of protection, *Decrease*. Before scheduling a job to be done from a scaffold, consider other means of access with less exposure to falls of personnel and objects. Boom trucks, scissor lifts, and rope access may be safer choices for short, easy jobs at lower elevations. These options have significant risks of their own, which must be managed well. Work planners and work crews should evaluate the entire risk profile before choosing a means of access.

Scaffolding is usually the best solution for large, difficult scopes of work concentrated in one area. Plan to move the materials to the worksite in bulk as much as possible using crane baskets or construction elevators. At some facilities, scaffolders pass every component by hand to dizzying heights, creating many opportunities for a tube or plank to fall.

Manual handling is unavoidable as the components are assembled. Most scaffolding crews use a silent means of communicating safe passage of materials from one person to another. One example is *twist and release,* where a quick twist of a pole signals that the receiver has control. The scaffolder below should stay out of the direct line of fire by standing to the side while handing off materials. Helpers should be outside the barricade unless performing a specific task and always out of the line of fire.

Scaffolds are just a means to execute other scopes of work. Scaffolders enable other workers to prevent dropped objects with solid construction, secure toe boards, and minimal gaps. If odd shapes or sizes create gaps, they should be covered. Where work or heavy foot traffic is expected, cover the sides from toe board to top rail with netting or alternative materials.

Summary

- Think broadly about dropped object scenarios when planning a job.
- Use the Dropped Object Risk Calculator to estimate the risk.
- Consider four layers of protection (**DROP**) against dropped objects from elevated work: DECREASE the amount of work at elevation as much as feasible, **RE**STRAIN any loose tools, materials, and equipment used

at elevation, **O**BSTRUCT the falls of any objects beyond the immediate work area, and **P**REVENT people from being in the line of fire.

- These controls can and will fail. Plan to fail safely.
- The higher the risk of dropped objects, the more layers of protection should be used.
- *Decrease*, or elimination, is the most effective control. The closer the controls are to the object, the more effective they typically are.
- The quantity AND quality of the controls matter.
- Barricades or exclusion zones are administrative controls and are moderately effective. For higher risk scenarios, bolster them.

Personal reflection and learning

- What are the most commonly dropped objects from elevated work at your site? Which activities produce them?
- How well did you identify the controls for the five scenarios? What did you learn?
- Visit the worksite. Which layers of protection do you see at the front-line? How effective are they? Which one needs the most improvement?
- Visit the worksite. How is housekeeping? What did you pick up? Plot some of the larger items on the risk calculator. What improvement tactic might work best at your site?
- Which additional controls might lead to the most improvement at your site?
- In your most recent and most severe dropped object incidents, which controls were present and effective? Which were absent or ineffective?
- How do front-line workers know which controls to use every day?
- What specific methods does your organization use to reduce the exposure to dropped objects from elevated work? What others might add value?

6

Dropped Objects from Elevated Equipment

If the incidents I shared so far scare you, this type may be terrifying. You may know this type of incident will occur, but not what, where, or when. There is little opportunity to *Restrain, Obstruct,* or *Prevent.* Barricades are rarely in place, unless an observant worker noticed something loose and reported it.

The fallen windsock described earlier is an example of how unpredictable this type of dropped object is.

> The purpose of the windsock is to show operators the wind direction in an emergency where evacuation is needed. During the previous turnaround, all windsocks were replaced. They were all rusted, and a few didn't spin. A few weeks later, one crashed to the ground. No one expected it since the windsock was new. We were lucky that no one was hurt.

> An investigation revealed that the technicians installed the windsock incorrectly, only partially threading the end of the pole into the fitting on the guardrail. The threads on the male and

female ends were mismatched, so the technicians could only screw them together for a couple of threads. Perhaps the work crew went to another high priority job and forgot to return to correct the problem. The windsock was going to fall sooner or later. It was that simple, but the fix is not so simple.

Controls for dropped objects from elevated work are implemented by a single work crew as the work is being done. The controls for dropped objects from elevated equipment are applied throughout the life of the equipment by different people from different departments. Equipment must be specified and designed correctly (engineers), installed properly and reliably secured (construction or maintenance technicians), and inspected and maintained over its life (inspectors, operators, and maintenance technicians). If the mounting of a light fixture on the ceiling of a warehouse or near the top of the distillation column deteriorates, the exposed personnel will be unaware of the risk and cannot protect themselves.

The controls for this type of dropped object are similar to those used in asset integrity programs intended to prevent major accidents from equipment failure and process safety events. Those events can cause even more tragic consequences than dropped objects. Asset integrity programs are extensive, and the controls for dropped objects will compete for similar resources.

Common falling objects

The most common falling objects of this type vary by facility. The following list pertains to a wide range of facilities. The actual list is endless. If it is mounted at height, it has fallen somewhere.

- Light fixtures
- Cameras, sun visors
- Pipe supports
- Cable supports, cable tray components

- Structural bolts, nuts
- Signs
- Insulation materials, including concrete fireproofing materials
- Out of service equipment
- Windsocks
- Speakers, alarm beacons
- Doors and components
- Ceiling tiles, sheetrock
- Building facade components
- Trees and branches
- Ladder components, swing gates
- Grating
- Semi-permanent scaffolds
- Electronic equipment
- Valve handles
- Hoists, sheaves, and other lifting equipment
- Crane boom cradle pads
- Masonry products

Common causes

In order to prevent these falling objects, the causes must be understood. Sometimes multiple causes are at work. For instance, corrosion may weaken a support for a camera, which is then knocked down by a windstorm. Vibration may loosen a gauge that a nearby scaffolder bumps with his elbow. Investigations should identify the root cause to enable robust corrective actions. If most of your incidents have a single cause, then one solution could lead to significant improvement.

Poor design

Causes include:

- Equipment not designed for the forces it will be subject to
- Incorrect materials of construction
- Equipment motion outside the design range
- Unnecessary equipment at height
- No Reliable Securing devices

Poor design was a contributing cause of the falling windsock. A securing mechanism was not in place to prevent components from falling if one of them failed.

Errors in design can lead to most of the other causes listed below.

Incorrect installation

Causes include:

- Substitution with incorrect materials
- Defective or incorrect equipment delivered
- Deviation from manufacturer's installation instructions
- Fasteners not fully tightened
- Poor welding

This was a contributing cause of the falling windsock. The supplied fasteners were incorrect and could not be fully tightened.

Construction and installation assurance processes may detect some, but not all, of these errors. Smaller accessories, such as windsocks and light fixtures, are not typically inspected as thoroughly as structural welds, high voltage line terminations, and heavy lift equipment. This is one of many reasons ongoing inspection programs are necessary. If the initial inspection does not discover a defect, perhaps the next one will.

Corrosion

Causes include:

- Dissimilar metals, leading to galvanic corrosion
- Harsh environments, e.g., sea spray for offshore and shore-side facilities
- Corrosive chemicals
- Inadequate painting or coating

A good understanding of the micro- and macro-environment of the equipment is critical for correct material specifications.

Vibration

Causes include:

- Nearby rotating equipment, such as pumps and compressors
- Fluid flow dynamics
- Wind effects

Vibration can damage and loosen pipe supports, valve handles, gauges, and other nearby accessories.

This phenomenon sometimes requires significant engineering analysis to eliminate. Large pumps, compressors, and similar equipment often need detailed testing after installation. Vibration can also occur as equipment ages and wears.

Collision

Causes include:

- Collisions with lifting equipment or loads
- Contact by workers performing a nearby task
- Strikes by another falling object

Light fixtures are a common target since many are mounted on elevated poles which can encroach upon lifting paths.

Ensure laydown areas are large enough to maneuver loads and minimize accessories protruding along common lift paths.

Before starting elevated work, workers should inspect the area for loose equipment. A little bump from a hip, wrench, or scaffold plank can turn a loose item into a falling object

Water damage

Water damage is a common cause of falling ceiling panels and building facades. Leaking roofs lead to many of these problems. Inspect them regularly and respond quickly to reports of water damage.

Mother nature

Though not typically the root cause, high winds can give the final nudge that knocks down a loose piece of equipment. These wind events range from short, intense wind storms to major events, such as hurricanes. Inspect for loose items before and after forecasted high wind events. In addition, consistent winds of lower speed can cause light poles and other extended equipment to vibrate and work loose.

All the causes described so far act upon equipment. However, ice and trees are falling object hazards in themselves. Dead or weakened trees usually fall during high wind events, but can fall at any time. Routine tree care can reduce the probability and is especially important near walkways, roads, parking areas, and occupied buildings.

In northern climates (in the northern hemisphere), ice is a troublesome dropped object risk. Changes to equipment or building drainage may help reduce ice buildup, but don't eliminate it. Therefore, access control, deflection shields, and controlled removal are also used.

Controls

The primary controls used to prevent dropped objects from elevated equipment are:

- Design and specification, including Reliable Securing
- Installation assurance
- Inspection
- Maintenance and repair

Reliable Securing is really a part of design, but it will be described in its own section due to its prominence as a dropped object prevention control. Even though part of design, it is often applied after the fact, as a result of dropped object surveys.

Design and specification

The causes described above highlight the importance of design and specification in preventing dropped objects. This control is similar to the *Decrease* control for elevated work, i.e. eliminate the risk factors causing equipment to fall.

Engineers are human. I know it's hard to believe, but they make mistakes too. Therefore, systems are in place to ensure their designs and specifications are correct:

- Engineering standards
- Competence assurance to verify that engineers have the right knowledge, skills, and experience to do their work–and are working within their defined roles
- Technical assurance, or layers of technical reviews, by staff with more expertise to ensure the risks are identified and the correct standards and specifications are used
- Original Equipment Manufacturer (OEM) expertise. The OEM knows the equipment best. They learn of incidents related to their equipment from a variety of customers and environments. The better the OEM understands how and where the equipment will be used, the better their support.

Reliable securing

Reliable Securing is a term that originated in the oil and gas drilling industry. DROPS (http://www.dropsonline.org/) has created a guidance document with the same title.[6] Reliable Securing aims to prevent many of the causes identified earlier (corrosion, vibration, wind events) from leading to dropped object and provides a fail-safe mechanism in case the fasteners fail. The three components of Reliable Securing are:

- Primary Fixing
- Secondary Retention
- Safety Securing

The focus here is on fixtures and accessories, not major structural components. Structural components should be engineered, inspected upon installation, and subject to a rigorous, ongoing inspection and maintenance program. The consequences of failure of these components can be more serious than the worst dropped fixtures and accessories.

The Reliable Securing guidance document contains over one hundred pages of examples, descriptions, and photos of methods to secure many kinds of elevated fixtures and accessories. It also describes some of the controls for dropped objects from elevated work and material movement. The document is a fantastic resource, so I will not replicate its content here. Instead, I will introduce the concepts and show how they fit into the larger dropped object prevention effort. Some equipment covered in the guideline is specific to the oil and gas drilling industry, but the guideline also addresses more common equipment. Free copies are available on the DROPS website (http://www.dr opsonline.org/). Download a copy and become familiar with it.

Primary fixing is "the primary method by which an item is installed, mounted and secured to prevent the item falling, (e.g., bolted connections, screws, pins, buckles, clips, welds etc.)."[7]

Welding and bolting are the most common primary fixing methods. Since some methods are more robust than others, the method used will determine the other layers of protection needed. Generally, the equipment supplier will specify the primary fixing method. The buyer should verify compatibility with the environment where they will install the equipment. A wrong choice of metallurgy can lead to corrosion and failure.

Secondary retention is "the engineered method for securing the primary fixing to prevent loss of clamping force or displacement of fastening components, (e.g., locking washers, locking wire, split pins / cotter pins, etc.)."[8]

A lock nut or lock washer reduces the likelihood that vibration will work a nut loose. Cotter pins keep lifting gear, such as shackles, from working loose. The rigger should inspect the shackles before every lift, but another simple layer of protection is worthwhile.

> A rigger found a large bolt laying on the deck of an offshore supply vessel. By large, I mean twelve inches long and two inches in diameter. Where did it come from? No equipment was located above the area. Perhaps it fell off a piece of equipment already delivered? That would lead to a lot of work. They would need to contact the facilities to which they had already delivered equipment to make sure the equipment was not missing a bolt. But no one recognized it as the type of bolt used on the offloaded equipment. A hunt for the source of the bolt ensued. They finally noticed that the base of the crane was missing one of a dozen such bolts. The bolt must have worked loose, fallen to the deck, and rolled over to the laydown area, perhaps during rough weather. Since eleven of twelve bolts remained, the crane still operated normally due to the engineering margin of safety.
>
> According to the DROPS risk calculator, a fatal injury could have occurred if someone had been in the wrong place at the wrong time. The area was not typically barricaded since no load could pass overhead so close to the base of the crane. If you refer to the Reliable Securing document, you will find this specific scenario, since cranes are common on drilling rigs. The vessel owner chose to prevent recurrence by threading wire through the twelve bolt heads so no single bolt could come free. A bolt might work loose, but not enough to fall. A photo of this solution is in the Reliable Securing guideline.

Safety securing is "an additional mechanism for securing the item to the main structure, suitably selected to restrain the item or its components from falling should the primary fixing fail, (e.g., rated steel or synthetic nets, lanyards, baskets, wires, slings, chains etc.)."[9]

This is similar to the *Obstruct* control for elevated work, a fail-safe control, i.e. if something starts to fall, don't let it fall far enough to hurt anyone. For example, a wire mesh net can be placed around a light fixture. If the fastenings work loose because of corrosion, the net will catch the fixture. If a load on the crane bumps a fixture, breaking it into pieces, the net will contain the larger pieces. Another example is a wire rope tether attached to a pulley.

Going back to the windsock story, how did we ensure the incident would not reoccur? The primary fixing was a screwed connection between the windsock pole and a socket on the guardrail. The connection was not made correctly, and no secondary retention or safety securing was present. Windsocks must be able to rotate, so a creative design was needed. During the investigation, we found this same incident had occurred many other times, but the design engineers were not aware.

The supplier was asked to provide options. They were eager to assist and developed a lanyard system that passed through the pole, connecting the major pieces. If any piece became loose, the lanyard would prevent the piece from falling further. All the windsocks were replaced with the more robust design. The supplier soon made it their standard design for everyone. This type of cooperative learning helps everyone become safer.

Installation assurance

Staying with the windsock example, how does one ensure that accessories at height are installed per manufacturer's recommendations and consistent with Reliable Securing principles? The range of options includes:

- Formal verification of asset integrity
- Routine installation assurance processes
- Ongoing inspections

Formal verification of asset integrity is generally reserved for *safety critical equipment,* which can cause catastrophic events if there is a failure. The scope of this formal verification program is enormous at large facilities processing hazardous materials and includes specific inspection and testing protocols, qualified inspectors, and records of the inspections. It might be applied to very large overhead equipment, but will not generally be applied to smaller accessories such as windsocks, light fixtures, and security cameras. There may be tens of thousands of such accessories in a very large facility.

Routine assurance of proper installation can be done as an integral part of the construction or installation process. It includes verification that the delivered equipment meets the specifications and is undamaged and verification that the equipment is installed per OEM instructions. For complex and higher risk equipment, inspections are done by a mix of OEM representatives and Company inspectors. Smaller, off-the-shelf equipment can sometimes slip through the cracks.

Finally, facilities using an ongoing dropped object survey will have multiple opportunities to detect installation errors or missing safety securing. Knowing that installation assurance processes are not perfect and may only be applied to a portion of the equipment reinforces the importance of using such surveys for the life of a facility.

Inspection

Ongoing inspection of elevated equipment is an essential control for preventing their falls to lower levels. Design is also essential, but even if that is done perfectly, a piece of equipment may become a fall risk after being bumped by a crane load or subjected to changing environmental conditions.

In many cases, ongoing inspections can detect such deterioration before equipment falls.

I had the opportunity to review the dropped object prevention program at a power generation facility. While on my first site walk, I kept lagging behind my escort, especially when climbing stairs or walking on elevated decks. My escort noticed that I was looking closely at signs, light fixtures, and cable supports, and occasionally tugging on the mountings. He eventually became frustrated enough to ask, "Why are you looking so closely at that small equipment? I thought you would be more interested in seeing scaffolding work or other work at height."

"Yes, of course. Let's continue to head in that direction. I am checking how well some of these accessories are mounted. You would be surprised how often these things work free and fall."

"OK. Now that you mention it, a couple of our light fixtures have fallen over the last year, but we never really found the root cause. Maybe you can help us."

At the end of the day, I met with my escort and sponsor to review key observations and the plan for the next day.

I kicked off the conversation. "Elaine was an excellent escort today. She showed me a lot of work at elevation and tolerated my observations of accessories mounted at height. Speaking of which, I would like to speak to an inspection supervisor tomorrow, if possible."

Marianne, the safety manager, said, "Thanks Elaine. I knew you would do a good job. Arnold, tell me more about these mounted accessories. I'm not sure why that is part of your review. I thought you were here to look at our work at height, scaffolding, and lifting?"

"Yes, I am interested in all of those things. They can all create dropped objects. I understand that a couple of light fixtures have fallen from high elevations, and I noticed several fasteners on other accessories were rusted or loose. I just want to understand how the site monitors these potential dropped objects."

"OK. I don't think we have a specific program for that," Marianne said as she raised her eyebrows. "Elaine, can you set that up for him? I'm not sure what he will find out."

The next morning, on my second site walk, Elaine and I stopped by the control room to notify them of our tour. I approached one of the operators and introduced myself. "You should be proud of your housekeeping. I noticed it was quite good on our initial walk yesterday. I have a question for you about something else. How do you monitor light fixtures, cameras, and other accessories to make sure they don't become loose and fall?"

"Each operator performs his or her rounds every shift to check for any leaks, smells, or unusual noises. We would notice and report any loose equipment if it was out there."

My interview with a mechanical inspector was later that afternoon. "Hello, thanks for making time for me on short notice. I presume you heard why I am here?"

"Yes," said Daniel, the mechanical inspection supervisor. "I don't know why you want to talk to me about dropped objects, but we're here together now, so let's do it."

"My primary questions are how does the site ensure that accessories mounted at height, like light fixtures, instruments, pipe and cable supports, etc. don't fall and what is your team's role in those efforts?"

"You'll have to talk to the electrical inspection supervisor about light fixtures and cable supports, but we don't have a formal program for that. I'm not sure we need one. My inspectors check things like instruments, valve handles, and pipe supports while they are performing our other required inspections. If something is loose or deteriorating to the point where it may fall, they will note it and report it to maintenance for repair."

"On my site walk yesterday, I noticed that the mountings on several accessories were severely corroded and a few were loose."

"We do the best we can. My inspectors aren't perfect, and they have a lot of other things to look for, like corrosion of piping and excessive vibration."

"OK, thanks. I'll try to find the electrical inspection supervisor. If you have the time, perhaps you can come to my report-out on Friday. I'll try to have some recommendations that will help reduce this risk of dropped objects."

Of course, these specialist inspectors would catch some of the problems related to dropped objects, but they would not catch all of them. Most wouldn't even look at all of them, since they had many other risk factors they were looking for. When I reviewed the inspection protocols (instructions on what to inspect and how), I found nothing related to the fastening of these smaller accessories.

Focused and systematic

If you look for everything, you will find nothing. OK, maybe a little–but far from everything.

DROPS provides excellent resources on an inspection program to prevent dropped objects from elevated equipment. It is used extensively in the oil and gas drilling industry. Some industries may struggle to implement it as is, but the inspection program has two critical success factors that should be maintained. The program should be **focused** and **systematic**.

To find specific problems consistently, an inspection should be focused on that problem. If an inspector is spot checking wall thicknesses on piping, they may not see a loose cable tray six feet to the left. That inspector has a lot of pipe to inspect and is looking for a very specific, and important, defect. If we are going to find the loose cable trays or incorrectly installed windsocks, someone must be looking at them specifically. This is **focus**.

Systematic means a well-structured and documented program designed to inspect all the objects in whatever scope you define. The scope may start small, but within that scope, inspect everything. If you don't want any more surveillance cameras to fall, make a list of their locations and create a schedule to inspect them all.

Picture books

Many years ago, the oil and gas drilling industry established a focused and systematic process for preventing dropped objects from elevated equipment. In their terminology, they are known as static dropped objects. Their facilities are smaller than most manufacturing facilities, but their dropped object risk is high due to large differences in height and extensive material handling by overhead equipment. Again, just the key aspects will be covered here. DROPS provides more detailed information in their guidelines, including a template for a dropped object inspection survey, or picture book.[10]

When a new drilling rig is built, and then every three (or so) years thereafter, third-party specialists inventory and inspect every non-structural piece of equipment that could fall, everything from lifting gear to gas detectors to grating. Some or all of this crew is trained in rope access since they cannot inspect all equipment from existing access routes.

For each piece of equipment, they take a photo and describe the equipment, location, type of fastening, recommended inspection frequency, and what the inspector should look for. They capture the information on a spreadsheet, an example of which can be found at the DROPS website. This inventory is the picture book or dropped objects survey. The survey can also be documented in an online database.

The inspectors enter any deficiencies in securing devices or deterioration of the fastenings in a repair log. For instance, they note equipment in need of safety securing. If a fastening is missing a nut, they note it. If the risk of a fall is imminent, supervisors block access to the area until they can remove or repair the equipment.

Between these third-party inspections, the operations and maintenance personnel divide the inventory into smaller pieces and inspect as best they can from existing work platforms. They have baseline photos and descriptions to help identify any changes since the last inspection. At more proactive sites, if someone is using a scaffold or rope access for another task, they take the relevant portion of the survey and inspect all nearby equipment while there. Why waste a great opportunity?

Whether done by specialist inspectors or company staff, they should inspect the equipment thoroughly.

> I read an incident report describing a large bolt falling eighty feet from the derrick to the rig floor. The drilling company completed their dropped object inspections as outlined above, but the investigation showed that the bolt had worked loose over time. The previous inspection reports showed no anomalies. Inspecting bolts in a derrick, grating clips, or light fixture mounts can be repetitive and uneventful, but is so important in preventing dropped objects. Most will be OK or there would be many more dropped objects. But every item must be inspected thoroughly to find the few exceptions that lead to danger. Apparently, that one bolt was missed or glossed over. How many others were out there?

A risk-based strategy

"Are you saying we should inspect every light fixture, alarm beacon, and sign every year? You've got to be crazy! We have thousands of light fixtures alone."

I was presenting the results of my audit at a large plant, and the audience was becoming rowdy. I wasn't surprised. I had seen the raised eyebrows and shaking heads on other site visits when I shared how drilling rigs inspect for potential dropped objects. But this was the most direct challenge I ever received.

I was not recommending this approach as is. Instead, I was sharing a proven practice some oil and gas drilling companies have used to reduce dropped objects by ninety percent over the last decade. I realized long ago this specific approach may be infeasible for large and complex facilities. This facility was at least one hundred times as large as a single drilling rig, and it inspected thousands of items for other hazards, some even more important than dropped objects. I introduced the practice as a concept from which they could develop a risk-based solution.

Some leaders react to discussions like this by dismissing the idea and changing nothing. Of course, we all know what that leads to–the **same result**. But endless variations exist between the two extremes. You may not be able to implement a full inspection program as described by the DROPS guidelines instantaneously, but you can start with a risk-based approach. This reaps the biggest benefit first without overwhelming your organization. You will continue to be vulnerable to falls of objects which are not being inspected, so continue to improve and expand the program regularly.

Here are some criteria which may be useful in developing a risk-based program to start the improvement journey:

- Type of equipment: Find out what falls at your location and what works loose. What objects fall at similar facilities in your company and industry? Light fixtures are light fixtures. Make a list of the top ten and start at the top. Inspect all light fixtures in the first quarter, then move to the ceiling tiles, and down the list. When finished with the list, recycle it or add other types of equipment. With this approach, the known problems are addressed first.
- Age: If the facility has been modified or expanded over time, focusing on the older sections first may be a good approach. Some risk factors, such as corrosion and vibration, worsen over time.
- Environmental conditions: Parts of the facility may face the brunt of the wind or sea spray. Other areas may be exposed to fugitive releases of corrosive chemicals. Winds may howl through narrow alleys. Make a list, prioritize it, and inspect those areas first and most often. This can be supplemented with one-off inspections after significant storms.
- High vibration: Areas of high vibration are usually known because of the noise created and high maintenance needs. These areas could be the focus while trying to mitigate the source of vibration. This can be a long

and difficult process, so a focused inspection program can reduce the risk in the meantime.

- <u>Differences in height</u>: Since the distance an object falls affects its level of risk, inspections could be focused on areas with large differences in height. A good example is the perimeter around a large process unit, conveyor, or building. You can combine this with the next factor to make it even more risk-based.
- <u>High-traffic areas</u>: The more people present, the higher the risk. Examples include internal roadways, designated walkways, lunch areas, tool storage areas, material staging areas, etc.
- <u>Area</u>: Perhaps dropped objects are more prevalent in one area of the facility for unknown reasons. Maybe the maintenance hasn't been as good as elsewhere. Or perhaps another contractor designed and constructed that portion of the site. These differences can have a large impact over time. This may be a good option when none of the other criteria make sense. It might not be risk-based, but at least it provides a way to organize the program. Pick an area of the facility and try it out. A small start is better than nothing, and will provide invaluable lessons to expand the program.

Whatever criteria you use to develop your program, the key elements of such a program are:

- Document the program. Include the risk-based criteria used, the scope, the inspection frequency, how to apply the Reliable Securing guidelines, when to review the program for expansion, how to measure progress, how to document the results, etc.
- Create an inspection schedule and stick to it.
- Assign inspectors. Show them what to inspect and what good looks like. For instance, which light fixtures should have safety securing? All of them, those with potential fall distances over fifty feet, those near common lift paths? Don't leave it to the individual inspectors. You owe

them more clarity. Otherwise, you will lose consistency and credibility.

- Document the inspections. Is a picture book necessary? It helps, but you might start simpler. The important first steps are the inspections and repairs. You may decide to document the results more completely later.
- Include the repairs in your maintenance management system so they are considered alongside the other needs. Apply the DROPS risk calculator to generate consistent risk priorities. Establish target times for the repairs. If loose accessories are not repaired or removed for months, how closely will your inspectors check next time? What message are you sending? Maybe that's why that bolt fell from the derrick?
- Establish key performance indicators. Examples include: number of inspections overdue, number of corrective actions outstanding, number and/or weight of potential dropped objects eliminated. Measure progress, identify trends, and take action.

Incorporating potential dropped objects into existing inspections

"Why do I need a separate inspection program for dropped objects? Can I incorporate this into my existing inspections? I have so many inspections already!"

Yes, it is possible. But it should still be systematic and focused. By definition, including these dropped object inspections into other inspection programs makes them less focused. However, integrating the two should be more efficient. In order to be successful, consider the following:

- Incorporate specific inspection points related to fasteners and dropped object risk factors in the inspection protocols for other equipment. In addition to checking the wiring on the light fixture, the inspector needs to check the fasteners and note the condition in the inspection report. While looking for corrosion under insulation on piping, the inspector also needs to inspect the pipe supports, instruments, valve handles, and other accessories.

- Inspection periods may need to be shortened. A ten year structural inspection frequency will be too long for accessories.
- Some isolated equipment may not fit within existing inspections. Separate protocols may be needed for them.
- Inspectors should be trained on dropped object risk factors, Reliable Securing, and site-specific guidelines.

Maintenance and repair

Deficiencies observed during inspections, operator rounds, or other observations should be entered in the facility's maintenance management system (MMS) so they can be tracked and repaired. One challenge in managing this risk is the number of competing repairs, many of which are needed to maintain production. There is always a backlog, the bottom of which can take a long time to be fixed.

The better maintenance management systems use risk assessment to help prioritize repairs. Safety risk should have a strong influence on the priorities if leadership is truly committed to *no harm*. Many sites underestimate the risk of dropped objects in such systems. It's usually not deliberate, just a lack of understanding of the risk. Remember the DROPS risk calculator–it can help here. If something is in the red area of the calculator, prioritize it as a potential fatality versus a minor infrastructure issue. That should place it near the top of the queue.

This section has focused on corrective maintenance, but preventative maintenance activities recommended by the manufacturer or driven by local experience should be included in your MMS as well. This can prevent some of the causes listed earlier from manifesting themselves, for example, vibration monitoring and painting.

Summary

- Dropped objects from elevated equipment are an asset integrity problem.
- A wide variety of equipment can fall.
- Common risk factors include corrosion, vibration, weather, and incorrect installation.
- The most effective controls are design and installation assurance and Reliable Securing.
- Inspection, maintenance, and repair are other necessary controls.
- Inspection programs should be focused and systematic.
- DROPS (http://www.dropsonline.org/) has developed guidance and templates for Reliable Securing and inspection.
- For a large, complex facility, a risk-based program may be a good place to start.
- Don't be intimidated. If you don't start, you will get the same outcome or worse.

Personal reflection and learning

- What types of equipment fall most often at your site? In your industry? What are the typical causes?
- How effective is your overall asset integrity program? How well does it capture potential dropped objects?
- Which criteria would be the best for structuring a risk-based inspection program at your site? Why?
- Visit the worksite. What loose or deteriorating objects at height do you see? What is the cause of the deterioration? Report them to the Maintenance department. What examples of Reliable Securing do you observe? Which accessories should have safety securing?
- Talk to some of your inspectors. How do they inspect for potential dropped objects? How confident are they about finding all of them? Check a few inspection protocols for inspection points related to dropped objects.

7

Dropped Objects from Material Movement

Martin, the warehouse supervisor, pointed to the entrance gate where a flatbed truck driver was showing her delivery paperwork to the guard. "There's the delivery the Maintenance manager has been calling about for the last two days," he said quickly to the crane operator standing at his side. Martin had called for the crane to be on standby once he heard the truck was an hour out.

"I'll warm up the crane and call the rigger," Jan, the crane operator, replied.

"Great, we'll unload the twelve-inch valve first and take it to the tank farm right away. Just put it on this trailer. Then we'll unload the rest of the material."

Jan headed toward the crane cab, and Martin directed the truck driver to park within easy reach of the waiting crane. The driver climbed out of the cab and approached him. "Hi. I'm Claudia. I see

102

you are ready to go. You must really need this stuff!"

"Like yesterday, but I'm glad you're here. I'm Martin, and Jan is in the crane. We'd like to unload the valve first, since that is what Maintenance has been waiting for." Jan gave her a short wave from the crane cab, and Claudia responded with a quick nod.

As Claudia climbed on the trailer to remove the securing straps on the valve and the pipe spools, Ben, the rigger, grabbed the soft slings Jan had laid near the crane. He joined Claudia on the truck bed and began wrapping a sling around the valve.

The four of them gathered for a quick toolbox talk. The crane operator led the discussion. "OK, we're going to unload the valve first, put it on this trailer, and then we'll unload the rest. Ben, don't forget to attach a tagline to the valve." Ben nodded and walked over to his cart to grab some rope.

"Aren't you going to barricade the area?" asked Claudia.

"We're kind of in a hurry and no one else is around, but sure. Martin, could you grab some cones from the warehouse? I have some red tape in the crane cab if you don't have any."

"OK, I have both inside." Martin hustled over to the warehouse.

After Martin erected the barricade, Jan moved the crane hook over the valve. Ben connected it to the sling, climbed down from the truck bed, and then backed away with the tagline in hand. He saw Martin standing just inside the barricade, out of the way. "Good enough," he thought.

As Jan began to lift the valve, it rolled toward one of the pipe spools before coming up off the bed of the trailer. The adjacent pipe spool also began to roll. Martin saw that the sling had snagged the pipe spool. He yelled to Jan to stop as he ran toward the truck. Ben tried to grab Martin's shirt as he sprinted by, just as the pipe spool fell off the truck and landed four feet in front of Martin.

Ben dropped the tagline, ran to Martin, and wrapped his left arm around him. "I've got you. You were so lucky. What were you thinking? The barricade means stay out."

"I know. It was instinct. I was just trying to help." His voice was as shaky as his legs. "If I was still in high school track team shape, I might not be talking to you now."

Martin was lucky. Much worse outcomes have occurred elsewhere. People have lost not only their toes, but their lives.

Government regulations, industry standards, recommended practices, and books on these subjects abound. Use these materials to develop or improve procedures and training at the site. Rather than replicating the detailed content here, I will highlight some types of movements that create dropped objects and share principles that can reduce dropped objects.

The controls for elevated work are mostly behavioral; those for elevated equipment are mostly related to asset integrity. Both are critical for dropped objects from material movement.

Principles for preventing dropped objects from material movement

Plan ahead

When is this not a good principle? Planning is so important here because it enables many of the principles listed below, especially for larger activities. It provides time to evaluate the best type of lifting or transport equipment and make it available at the right time and place. It allows pre-staging of materials in bulk near the work sites. If space is tight, movements can be planned to reduce double or triple handling. Work crews can aggregate materials in containers and move them once. For example, some companies involved in offshore oil drilling collaborate with their suppliers to aggregate materials and equipment into larger, common containers versus having each supplier use their own.

Secure the load

Whatever the mode of movement, secure the load no matter how small the movement is.

- Follow regulations, industry standards, and best practices.
- Package smaller items in containers rated for lifting/moving.
- Check for loose items before moving.
- Strap loads to truck beds, trailers, pallets, even small carts.

Move in bulk when feasible

Moving smaller items in bulk can reduce exposure to dropped objects. This goes hand in hand with the next principle. With fewer movements, there is less opportunity for human error.

Mechanical over manual

Instead of taking five trips up three flights of stairs to bring tools and materials to the work site, can a crane be used? Perhaps different work teams can coordinate their needs to justify the special equipment. This has the added benefit of reducing the risk of slips, trips, and falls.

Equipment inspection and maintenance

Cranes, forklifts, and similar equipment are heavily utilized at many sites. Finding the time and space to inspect them can be difficult, but is essential. The main lifting gear will have an inspection and maintenance protocol and schedule, but inspections of smaller accessories, such as cameras, lights, and anemometers, can be overlooked. This equipment is moving all the time, so things can work loose quickly. Remember, accessories on crane booms are often very high in the air.

Operations procedures and competent operators

Ensure equipment operators are competent in operating the specific equipment being used and follow operating procedures. Use the equipment only as specified by the manufacturer.

Modes of movement

Now let's review common modes of movement that create dropped objects. I will not cover them in detail, just point out key tips related to dropped object prevention.

Lifting and hoisting

Lifting and hoisting can be a large contributor to dropped objects at sites with weak programs and high activity.

Controls to prevent dropped objects from lifting and hoisting include:

- Proper equipment selection
- Inspection and maintenance of the main lifting gear, lifting appliances, and crane accessories
- Competent operators and riggers
- Competent spotters or banksmen
- Lift plans
- Exclusion zones and/or signaling protocols

Competence of crane operators receives a lot of attention, but the competence of the riggers is also critical for success. Failure of rigging generally causes more dropped objects than failure of the main lifting gear. Lifting appliances should be inspected before every use by the rigger and periodically by a specialist. Many sites use a color-coding system for rigging inspections so riggers and others can quickly spot equipment with overdue inspections.

Detailed inspection of all parts of lifting equipment is important since some can operate at extraordinary height. A light fixture falling from an extended crane boom is dangerous.

Inspect loads for items that are loose or may become loose as the load flexes during movement. Examples include extra nuts and bolts, left-behind tools, rocks picked up from the laydown yard, and leftover sea-fastening materials.

When placing loads in tight spots, use spotters, tag lines, and/or push/pull poles to avoid bumping any fixtures or other nearby equipment. Riggers and spotters must stay out of the line of fire, which is often larger than the area directly underneath the load. Loads can tilt or be deflected when they fall. Riggers are not strong enough to overcome the massive forces associated with large loads anyway. Push/pull poles enable riggers to nudge the load or grab a tag line from a safe distance.

Consider safety securing for objects mounted at high elevation in common lift paths or near laydown areas, as discussed in Chapter 6.

Managing exclusion zones is more difficult for large or mobile cranes. Load travel paths can be long and vary from load to load, and cranes can move between loads. When cranes are stationary, conventional barricades may be an option. Otherwise, banksmen may use flags, whistles, horns, and other signaling devices to keep people out of the line of fire.

Overhead hoists typically receive less attention than cranes. Competent persons should design, test, inspect, and operate them.

Trucks

Potential dropped objects from trucks and other vehicles are often over-looked. A site may invest significant effort to reduce dropped objects from elevated work and lifting and hoisting, then be surprised by dropped objects from trucks or forklifts.

Load securement is a key control since the consequences of loads falling on public roads can be severe. Because of the public exposure, very detailed guidelines and regulations are available. This section will focus on loading and unloading at a facility.

Short trips within the facility may tempt drivers to take shortcuts on load securement, but a small pothole or a sharp turn may cause a dropped load.

Examples of dropped object scenarios when unloading or loading trucks include:

- Pushing an object off the back side when using a forklift
- Unexpected movement due to a release of tension while removing straps or chains from the load
- Snagging one object while moving an adjacent one
- Damaged containers or pallets

Exclusion zones are typically not used as often for truck loading/unloading as for elevated work or lifting and hoisting. The low height of the truck bed leads some to underestimate the risk. Exclusion zones should protect people from the movements of the crane or forklift and keep them from being in the line of fire of the load, including the back-side of the trailer. Riggers and spotters should stay out of the exclusion zone when not performing a necessary task and never enter the line of fire. Sudden movements can change lives in a heartbeat.

Forklifts

Load securement is a critical control here as well. Do you detect a common theme? Once again, detailed guidelines are available from manufacturers, industry groups, and regulatory agencies. Containers, pallets, and other lifting accessories must be in good condition and suitable for the load.

Forklifts can dart in all directions, and the operator may have limited visibility. The operator must watch the load, nearby personnel, and other vehicles. A spotter is essential for safe operations in congested areas.

Use forklifts only as recommended by the manufacturer. Experienced operators occasionally use these machines for creative and delicate tasks. They accomplish the job in the short term, but how do they affect the equipment over the longer term? Attachments are available to handle some awkward loads, such as pipes and drums.

In warehouses, sturdy containers/pallets, good organization, and housekeeping are essential to preventing loose items from falling off of shelving. Keep the riskier loads (heavy, awkward, open tops) closer to the ground when feasible.

Carts

Good practices include using closed carts, securing the loads, and covering guardrails in elevated high-traffic areas. Objects don't need to be on fast-moving, large vehicles to be dangerous.

> Workers were pulling a ninety-five-pound valve on a cart across the grating on a deck thirty feet above ground. The cart had a two-inch lip around the edge. The valve had no securing straps, but several protrusions kept it from rolling.
>
> "This looks like the wagon I used when I was five years old. What's the worst that could happen? If it tips over, the valve falls twelve inches?" When you were five years old, you did not use your wagon three stories up, and you pulled a one-pound teddy bear versus a ninety-five-pound valve.
>
> A wheel caught a piece of lumber laying near the walkway. Before the pipefitter could react, the cart tipped over and dumped the valve over the toe board of the nearby guardrail to the ground. An operator below crumbled to the ground in terror as it landed just five feet away. With foot traffic along the path growing, they installed wire fencing in the gap between the toe board and the handrail to prevent recurrence.

Containers

Whether using cranes, trucks, or forklifts, many materials will be in or on containers, ranging from sturdy steel boxes to wooden pallets. The load is only as strong as its container. If the container fails, both it and the contents are likely to fall. Containers should be properly rated, fit for the load and lifting device, and in good condition. Not sure–don't move it.

> I once saw a photo of a large steel plate, four feet by four feet and one-inch thick on the ground by a flatbed truck amongst a pile of splintered lumber. The investigation revealed that the pallet was oversized and made of poor quality wood. The forks on the forklift didn't reach the last couple of boards, and the pallet fell apart during the lift. Fortunately, no bystanders tried to save the day and put themselves in harm's way.

Consider the movements a piece of equipment or bundle of materials may need from initial receipt to its final destination. In some cases, containers and rigging may be designed to be used for multiple movements, thereby reducing the need to transfer materials in and out of containers multiple times. Of course, the container, rigging, and loads should be inspected prior to each movement.

Scaffolding

For facilities using extensive scaffolding, these components can be the most common objects dropped. Many of these occur while passing materials from one person to another. While manual handling is necessary during final assembly, it is rarely necessary when moving materials to higher staging areas. Use mechanical means to move the components in bulk when feasible.

A common method for moving scaffold components is the gin wheel. While better than manual handling, other risks remain: poor load securement, hitting obstacles while lifting, losing control of the rope, ground operators in the line of fire, poorly mounted wheels, and lifting bags that are too small. The National Access and Scaffolding Confederation (NASC) has published several guides on safe scaffolding practices including lifting of components.[11] This guide describes a hierarchy of lifting methods, generally from safest to riskiest:

1. Tower cranes
2. Mobile cranes
3. Forklift trucks
4. Passenger/goods hoists
5. Mechanical hoists
6. Gin wheels
7. Hand lines
8. Manual handling

Let's consider the difference in exposure between method 2 and method 8 in more detail. Let's say a scaffold is needed on the upper deck (fourth story) of a module. The scaffolders can ask a crane to lift all the materials (poles, planks, clamps, wire) to the upper deck in baskets and racks. How could components fall? The rigging could fail, the baskets may be overfilled, the basket may tilt when colliding with an obstacle or with jerky movements, or accessories could fall off the crane. Alternatively, the components could be moved to the upper deck by a human daisy chain. Let's say the decks are fifteen feet apart, for a total of forty-five feet. It might take five to ten people to move poles and planks that distance, depending upon how the access stairs are configured. Clamps may be carried up the stairs in bags. So we have hundreds to thousands of manual hand-offs, depending on the size of the scaffold. How could components fall? The scaffolder below releases his grip on a pole or plank before the one above has control, a component collides with the structure while being raised, a pole falls over the edge when placed in a temporary storage area, the bag of clamps is overfilled, the person carrying the bag of clamps trips going up the stairs, etc. Which scenario is more likely to create a dropped object? And to complete the risk picture, which scenario has more people near the line of fire?

While all methods listed above may not be feasible for all sites, the hierarchy can help you reduce risk. This hierarchy can also be applied to the movement of most other types of material. Notice the consistency with the principles identified earlier, namely moving materials in bulk using mechanical means.

Summary

- Move materials in bulk with mechanized means where feasible.
- Load securement is critical for all modes–many public resources are available to assist.
- Exclusion zones are more dynamic, but just as critical.
- Check for loose items before any movement.
- Care for the equipment, including inspection of accessories at height.

Personal reflection and learning

- Which mode of movement produces the most dangerous dropped objects at your site? Why?
- What opportunities exist for moving material in bulk by mechanized means vs. manually?
- Visit the worksite. How are scaffold materials moved to height? Discuss other options with the scaffolding crew. How well do they stay out of the line of fire?
- Visit the warehouse or other area with forklift and truck loading/unloading activity. How well are loads secured? How effective are the exclusion zones around trucks when being loaded or unloaded? When are spotters used for forklifts?

8

Put your Safety Management System to Work

The difference-maker

Having all the necessary tool lanyards, nets, barricades, dropped object inspections, and rigging at your facility is not enough to eliminate dropped objects. A good Safety Management System (SMS) must be in place to support it.

> I was the HSE manager on a multi-billion dollar project to build a new chemical plant. The production modules were being built in a fabrication yard in the Far East that was used on past projects. We knew dropped objects would be one of the top risks. Others included falls from height and confined space entry. Safety performance in previous projects was average, and of course, we aimed to be better, i.e. zero serious injuries. My team of safety coaches advised the contractor's supervisors and front-line workers on safe work practices.

Unlike previous projects, we developed a robust, dedicated plan to prevent dropped objects. At least, we thought it was robust. We trained the workforce, purchased lanyards of all sizes and types, draped netting on guardrails up and down the structures, and provided miles of red barricade tape for exclusion zones. I admit, we were confident!

In the first nine months, we had no serious dropped objects, further bolstering our confidence. But, the work was still near the ground. Then, in just one week, three objects with the potential to seriously injure someone fell out of the structures, including one that grazed the arm of a scaffolder. He was shaken, but not seriously injured, and the project team was scared. After the first incident, we started an investigation. Before we even finished it, we began another. After the third, the project manager called **time-out**, and we stopped work for several weeks.

When reflecting on the causes of these incidents, we realized that even though we had the materials and training to prevent dropped objects, the front-line workers were not implementing the controls consistently or effectively. Of course, they weren't the root cause. Over time, we (project and contractor leadership) had grown to tolerate uncovered gaps in work surfaces, three out of four tools being secured with lanyards, and guardrail netting fluttering in the breeze. A new norm was established, and we were paying the price. The exposure would increase as our construction moved higher.

We used the time-out for two things. First, we patched up the gaps on work platforms, purchased more tool lanyards, and established *no-go* zones in high-risk areas. But these quick fixes would not last unless we changed our systems. We decided to measure the effectiveness of our controls every day at the front-line, where it mattered most. Those at the front-line are exposed to the hazard the most, and they put the controls in place. I changed the role of my team from coaches to inspectors. They still advised and intervened, but they also observed and recorded the effectiveness of the controls at each worksite. Not after the intervention, but as they arrived at the worksite.

Using data on the condition of controls at the front-line to hold the leaders accountable made the difference, not the extra lanyards we bought or the pep talks by the big boss. The **system** change made it stick. We reduced the number of high risk dropped objects by over 80 percent in one year. Considering our previous rash of incidents, we may have saved a life or two.

This chapter will describe the components of an SMS which can strengthen the physical controls at the front-line. An SMS is a collection of principles, work processes, and tools working together to manage safety systematically. Many companies expand the management system to cover Health and Environment as well (an HSE MS). If you don't use an SMS, you can still use these tools and practices, but consider implementing a complete SMS for the maximum benefit.

The SMS structure in ISO 45001:2018 - Occupational Health and Safety Management Systems[12] is used to organize this chapter:

- Leadership and worker participation
- Planning
- Support
- Operations
- Performance evaluation
- Improvement

Leadership and worker participation

Leadership

It all starts at the top, but is so difficult to describe. A leader can reduce the risk of dropped objects in the following ways:

- Emphasize that harm from dropped objects is **preventable**.
- Provide the resources (supplies, people, time, training) to **do what you say**.
- Personally understand the specific dropped object risks and controls at the site.
- Check implementation at the front-line. Ask workers how they are managing the risk and what their challenges are. Don't walk by behaviors that don't meet site requirements.
- Measure leading and lagging indicators, with emphasis on leading. Review them frequently and act on the trends.
- Recognize outstanding performance.
- Learn from incidents and adjust the plan. If you are going to blame anyone, blame yourself. A little humility goes a long way.
- Focus on one area at a time. Get it right before shifting focus.

Most of these behaviors will help improve overall safety performance. If you need to improve on dropped objects specifically, focus is the key. In my experience, the benefits will spill over to other hazards. If workers know you care, understand the risk, are recognized for contributing to the solution, have the resources they need, and are held accountable for poor performance, the near misses and injuries should decline. But make sure you are truly committed to the cause. The organization is watching your every move, or lack thereof. If you are wondering whether preventing dropped objects is a priority, even without saying it, people will know.

"You get what you tolerate?" The saying is attributed to many people, underscoring its wide applicability.

> I once audited the dropped objects program at a large manufacturing facility. They had a detailed procedure, abundant supplies, and well-trained workers. From my preparations, this appeared to be an easy one.
>
> The initial interviews reinforced the pre-read, but when I went outside, only half of the workers used the controls. In fact, several had tool lanyards but were not using them. This observation did not surprise the site leaders. I was puzzled. They clearly wanted to reduce dropped objects as they were paying me to help them. But they did not act on the knowledge they already had. The workers had received a powerful, unspoken message, "It must be OK with the bosses, so why go through the extra hassle? It must not be so important after all."
>
> Leadership came so close to making a real difference, but they missed the last steps, follow-up and accountability. It takes more effort to undo this damage than to support the change from the beginning. "Trust us, we mean it this time," will not work in cases like this.

I never liked the saying, "What interests my boss fascinates me," but it does have a place where the boss' interest is compelling. Show interest, show commitment, show why it's compelling, and most importantly, show care.

Dropped object prevention culture

No one can anticipate every potential item that may fall. The list is endless, essentially everything above the ground. I audited dropped object prevention programs at many sites. Some sites knew they had a problem, but didn't know where to start. Some sites had begun, but were struggling with effective implementation. A few sites had implemented the basics well, but still had serious near misses, usually involving objects not expected to fall, for example, a long-forgotten section of out-of-service piping, an instrument on a large rental crane, part of a light pole, and yes, a windsock.

A few leaders shrugged these incidents off. "Those are one-offs. How could they be predicted? Our people always use lanyards and barricades." Other leaders realized these one-off incidents could seriously injure someone. "What else can we do? We can't put everything in a procedure. People won't read it or understand it, and we'll still miss things!"

Unfortunately, there is no simple solution to this dilemma. But while implementing the nuts and bolts of common controls, you can build a dropped object prevention culture, one where everyone asks three questions before and during their work:

- What could fall?
- Where could it go?
- How can I control the risk?

Not only when building a scaffold, replacing a pump on the upper deck, or lifting pipes from a supply boat, but also when unloading a truck, bringing a pile driver onsite, or moving a valve across the deck on a cart. Not only what could fall from their own work activities but also from their surroundings. If each individual or team looks at their own work or area of responsibility this way, the entire site can be covered. Dropped objects that are anticipated can be prevented. The massive scope of the problem is less intimidating when viewed this way.

Safety observation system

Safety observation systems for workforce participation are common. Some sites use formal Behavioral Based Safety (BBS) systems; others use less structured systems. These observation systems typically have forms to help guide the observations and discussion with the work crew and to capture the hazards observed and highlights from the discussion. The information captured on the forms is usually subjective, but it can be used to trend the hazards observed and to gather insights on potential improvement opportunities. If the hazard of dropped objects is included on the form, you are more likely to gain valuable insights from the front-line on the hazard.

DROPS committee

A DROPS Committee, consisting of Dropped Object focal points from various parts of the organization, can provide a conduit for direct communication with the leadership team. Consider including representatives from all organizations with a significant role in preventing dropped objects, e.g., Operations, Maintenance, Projects, Inspection, Logistics/Lifting, Warehousing, and Safety. Also consider representatives from contractors whose work is filled with dropped object exposure, e.g., scaffolding, mechanical, painting, electrical, rope access, and heavy equipment providers. Resist the temptation to fill the committee with safety professionals.

Invite the committee to share their unfiltered views on what is working, what needs to improve, and what help they need from you. This may be uncomfortable since you may need to provide additional resources or change work processes to maintain credibility, but the feedback is coming from those who know the work best. Consider it a gift.

Planning

Dropped object procedure

How will workers meet your expectations if you can't describe them in detail? Document specific requirements and guidelines in a dropped object prevention procedure or plan specific to the organization. An effective one will describe both the front-line controls and systems to be used, such as:

- Roles and responsibilities
- Contractor engagement
- Risk assessment
- Training and awareness
- Hierarchy of controls or layers of protection
- Front-line controls for all types of dropped objects
- Performance measurement and reporting
- Incident reporting and followup
- Auditing and review

In other words, it should cover the topics in this book. An example of a Table of Contents can be obtained via the download link in the Author's Note.

Some aspects may already be described in other documents. However, a separate dropped objects procedure that references them, includes detailed requirements not captured elsewhere, and shows how they fit together is a good practice. For instance, an existing Lifting and Hoisting procedure does not need to be replicated. Just provide a reference and explain how it fits into the overall dropped objects prevention program. Overhead equipment inspection requirements may fit in the site-wide inspection program and protocols.

Including dropped objects requirements for elevated work in the Working at Height procedure is tempting and logical, yet risky. Dropped objects occur in many scenarios where personal fall protection is not required. Workers may infer the dropped objects requirements are not applicable unless they are working where personal fall protection is required. Recall the story about Olga and "I'm not working at height."

Emphasize that accountability for preventing dropped objects rests with line management, not the safety department. The safety department has an important role to play in supporting leaders and the front-line; list those responsibilities as well. Initiatives driven by the safety department alone are rarely effective.

Dropped object risk assessment

Introduce the Dropped Object Risk Calculator and how it should be used.

Exposure reduction

Exposure reduction, the *Decrease* layer of protection, is by far the most effective control and eliminates opportunities for human error. But it doesn't just happen. Take the opportunity in your procedure and training to make this objective clear and to describe the specific, systematic methods used to reduce exposure at the site.

Contractor engagement

Influence the leaders of contractors to join the improvement journey. In many cases, their people are performing the work with the highest risk of dropped objects and are the most likely to be hurt. Without them on board, you won't get far. The procedure will be invaluable here. How can they commit without understanding what you want and why?

First, make sure they understand the risk to their people; their safety is why you are doing all of this in the first place. Second, ask them to develop their own dropped object prevention plan or procedure, consistent with yours and specific to their work scopes. For instance, procedures for scaffolding contractors can include details on how secure passage of materials to teammates is confirmed, how materials are transported to staging areas, etc.

This can be accomplished through informal influence or by including requirements in contracts. The latter is more robust, but the best approach depends upon the existing relationship with the contractors. In either case, hold them accountable for their performance.

Support

Training

Front-line workers need to understand specifically how they are expected to prevent dropped objects. Now you have a procedure, but they don't always read the procedures. They need to be trained on subjects such as:

- The risk
- Job-specific risk assessment
- What controls to use when
- How to use each control
- Where to obtain materials
- What to do if they can't meet expectations
- What to do if someone else isn't meeting expectations
- How to offer improvement ideas

A mix of classroom training and hands-on training is the most effective approach. Hands-on training can be done using a physical mockup of a scaffold or a permanent platform with grating, guardrails, and various types of gaps. Allow the workers to practice with lanyards, containers, netting, and materials for exclusion zones available at the site. Many specialist vendors will demonstrate how to use their products.

Front-line workers and their supervisors need detailed training. But this is not all on them; they need support. The rest of the workforce can benefit from awareness training, particularly anyone in the management chain of the front-line workers and those who visit the worksites frequently. They should understand the basics so they can recognize the risks, understand the basic controls, and discuss both with workers while in the field.

Awareness tools

Chapter 3 described a demonstration where tools and other materials are dropped on melons. The *melon drop* is one of the most impactful ways to educate the workforce about the risk of dropped objects.

> At another site, a much simpler technique made quite an impression on me. I was walking with the safety supervisor, Michelle, from the construction office to a ten story building under construction. We were still 200 feet from the building when I noticed several large signs on the side of the building with two numbers, one in black numerals with a white background, one with white numerals on a red background.
>
> "What are those?" I said, pointing to the highest sign on the ninth floor.
>
> "Oh that." Michelle said with a chuckle. "The *135* is the number of feet above ground level at that location, and the *0.75* represents the weight (in pounds) of an object that could cause a fatality if dropped from that location. As you can see, it doesn't take much at that height."

"Wow, that grabbed my attention. Much more so than the massive Stop Drops sign with the black circle and red slash. I can see the *feet* and *pounds* labels now that I look more closely. Who thought of that?"

"One of the safety inspectors, Shawn, from the prime contractor suggested it. He saw it used on a previous job. We gave it a try since it was easy and catchy. He just used the risk calculator to get the weights. We try to hang signs where most of the work is going to be for the next few weeks."

"I love it. It is very simple and caused us to talk about it before we even got to the building. If the workers are having those same discussions, you've got a hit here. Please introduce me to Shawn if we run into him on our site walk. I want to thank him."

Many other simple visual techniques have been used to reinforce the message. Memories will fade, new people will join the workforce, and other priorities will arise. Consider the following:

- Build a large bulletin board displaying the dropped object risk calculator near the entrance to the worksite. Attach photos of actual or potential dropped objects to communicate their risk level. Teams can use the board as a prop for Toolbox talks. The bulletin board can also include performance metrics and recognition, such as *days since last dropped object, effectiveness of controls over the last week,* and *work crew of the month.* Place a scale on an adjacent table so workers can weigh their tools and materials and plot them on the calculator. After doing this a few times, the workers will develop a better gut feel of the risk.
- Place a container near access points in which people can deposit loose items they find at height. Seeing this collection of items as they walk to work may impart a little guilt and motivate them to do a better job of housekeeping. Teams can take these items to the scale and bulletin

board during toolbox talks or safety meetings to understand the risk.

- Include the risk calculator on a pre-job risk assessment form or prompt card for use during toolbox talks to encourage discussion of the risk of the team's tools and materials.
- Share incidents in safety meetings and toolbox talks. Those from the site or company are the most impactful. When they are too far removed from your work, it's easy to say, "that wouldn't happen here."
- Recognize work crews with the best dropped object controls for the month. Buy them lunch or some fun work shirts they can wear with pride.

These are just a few examples. Use your imagination and encourage work teams to develop other innovative methods to build a dropped object prevention culture. I've seen work crews develop simple and effective solutions to the challenges they face in their own tasks. Recall those tarps with velcro straps. Another proactive crew used custom-cut plywood to cover piping penetrations in grating. If it doesn't happen naturally, sponsor a design competition for a control that needs improvement.

Supplies and materials

Dropped object prevention supplies should be available in convenient locations for the workers. Asking workers to apply netting to guardrails when no netting is available on the site is a sure way to lose credibility. In some cases, craftsmen have resorted to making their own tool lanyards from string, wire, or telephone cord. What a lost opportunity. Getting the commitment to use the controls is the hard part; don't stumble on the easy part.

For large sites, *DROP Boxes* or team tool boxes containing lanyards, netting, and tarps near the work site can facilitate implementation. Who wants to walk to the toolroom a quarter of a mile away when they tear their lanyard on a sharp edge?

DROPS focal points

Appoint a DROPS focal point who keeps these systems running and is always seeking lessons learned and better practices. This is not a full-time role, but give the focal point the time to make a difference. Typical responsibilities include:

- Maintaining the procedure or plan
- Chairing the DROPS Committee meetings
- Reviewing incidents for trends and valuable learnings
- Maintaining contact with vendors to discover useful products
- Developing and maintaining the training materials
- Reporting leading and lagging performance data
- Acting as a liaison between the front-line and senior management
- Participating in incident investigations
- Staying in touch with external developments

In a large organization, consider having another layer of DROPS focal points representing the various departments, trades, and contractors. These focal points can form a DROPS Committee, as discussed earlier.

Operations

Pre-job risk assessment

The dropped objects procedure should identify the type of pre-job risk assessment to be done before each job. An effective approach is described in Chapter 3. The risk assessment is the basis for choosing the right controls, so the workers must think broadly and get it right. Will the risk assessment be a stand-alone tool or incorporated into the standard tool used at the site? Will contractors use the site's assessment tool or their own? Will it be written or only discussed as part of the toolbox talk? Must a supervisor approve it before beginning work? How do the workers get help if the minimum controls are not feasible?

An example of such a tool can be found <u>here.</u>(https://strivingforsafety.maile rpage.com/downloads-dont-let-it-fall).

A risk assessment of dropped objects from elevated work may include questions such as:

- What objects could fall? At what height will you be working? What is the highest risk level?
- What will you do before going to height to reduce the risk of dropped objects?
- How will you restrain or contain the objects to prevent them from falling?
- How will you prevent objects from falling below your immediate work area? Which holes do you need to cover?
- How will you keep people below out of the line of fire? What type of exclusion zone will you use?
- Are you able to implement all the controls required by the site procedure? If not, what else will you do?
- Will other crews be working above or below you? Who has priority?
- How will you move materials to and from your work site?
- Are there any loose materials or mounted equipment near the work area? How will you prevent them from falling?
- How have you secured your personal accessories (e.g. hard hat, radio)?

Hazard hunts

Hazard Hunts focused on dropped objects build awareness, remove potential dropped objects from the workplace, and improve housekeeping.

Assign groups to different parts of the facility. Their sole task for thirty to ninety minutes is to find and remove (if safe) any potential dropped objects and place them in waste containers or closed bags. It is amazing how much can be found when only looking for potential dropped objects. The variety of objects collected will broaden everyone's view of what can fall. To make it more interesting, hide prize tickets in nooks and crannies where objects can hide for years. Prizes can also be awarded to the teams who remove the highest number or largest weight of objects.

> I recall walking up to the top deck of a module in a refinery and telling my escort, "John, the housekeeping is good."
>
> He gave me a quick nod in acknowledgement, but didn't smile as I expected. Maybe he felt this was normal, so a response wasn't needed. As I picked up enough bolts, grating clips, and small tools to fill my pockets and one hand, John shook his head vigorously as his chin sank.
>
> "Don't be ashamed. I am only looking for potential dropped objects, but you and your colleagues are looking for many other potential problems. Some of these things were hard to spot. Your housekeeping is better than at most sites I visit. Tell this story to your teammates at your safety meeting this week. They should be proud, but also motivated to find these hidden hazards."
>
> He stopped shaking his head and raised his chin, but still did not smile.

If you look for everything, you will find nothing.

Simultaneous operations (SIMOPS)

As I approach a work crew, I instinctively look up and down to see if they (and I) are at risk from dropped objects or creating a risk for others. This is an essential individual habit, but sites with high levels of activity should use a more formal process for managing simultaneous operations (SIMOPS).

This topic is covered in detail in Chapter 5.

Maintenance Management System

A Maintenance Management System (MMS) is a critical tool for managing the risks of dropped objects from elevated equipment and was covered in detail in Chapter 6.

Performance evaluation

Key performance indicators

"What gets measured gets improved," Robin Sharma said. Leading and lagging key performance indicators (KPI's) show how well your improvement efforts are working. Lagging indicators are based on undesirable events that occur. Examples include injuries from dropped objects, total dropped objects, and high potential dropped objects. Another example is the fraction of *Fail Safe* or *Fail Lucky* dropped objects, i.e. if they fell into an effective exclusion zone with no one present, they are *Fail Safe*. This measure also has attributes of a leading indicator, since it sheds light on the effectiveness of exclusion zones.

If trying to reduce high risk dropped objects, causes of lower risk dropped objects have attributes of leading and lagging indicators. I used this KPI for sites in my remit and noticed a large uptick in low-risk dropped objects at a smaller site. I found the lack of use of tool lanyards to be a common immediate cause. Since tool lanyards are also an essential control for more serious dropped objects, we found the root causes, took action, and nipped it in the bud. Lower risk incidents are tempting to ignore since they pose little threat individually, but tracking their numbers and causes as key performance indicators may reveal valuable insights on the health of controls used for preventing dropped objects more generally.

Lagging indicators are both helpful and necessary, but are not as powerful as leading indicators. Leading indicators can be predictive of future incidents. Examples related to dropped objects include:

- Degree of effective implementation of controls at the front-line, e.g., fraction of jobs with effective use of tool lanyards
- Fraction of jobs with a specific and complete pre-job risk assessment
- Fraction of elevated equipment inspections done as scheduled
- Fraction of dropped object-related repairs done within target time
- Number of falling objects from previously inspected areas. This is a measure of the quality of the elevated equipment inspections.
- Weight/number of potential dropped objects removed in Hazard Hunts. This generates a lot of interest, but is a low number good or bad? A low number may mean participants didn't look closely or there was nothing to find.
- Fraction of vehicles arriving/departing with secured loads
- Fraction of crane loads checked for loose items before lifting

Other indicators are further removed from the work, such as the number of workers trained on time and the number of awareness campaigns. These are easier to measure but not as effective because they don't show how well controls are implemented. Just because someone was trained doesn't mean they are applying the controls every day. In my opinion, the KPI's closest to the work are the best. Data collection requires more effort, but you get what you pay for. Consider both the value and effort when choosing a leading indicator. Indicators with low value may give you a false sense of security.

It is not necessary to track all the KPI's listed above; doing so would make it hard to figure out the real message. The examples above cover all three types of dropped objects, so you can select one or two relevant to your focus area. If trying to improve the controls for elevated work, the first one may be all you need. It is one of the best on the list and can be customized to any site by adjusting the questions.

When only measuring lagging indicators, the absence of dropped objects may lead you to believe the war is over. Or were you just lucky for a few months? Has your laser focus and response to previous incidents stifled reporting? If leading indicators are being monitored as well, they may answer these questions. Conversely, if the leading indicators are positive, but the lagging indicators are not improving for six to twelve months, you may have chosen the wrong leading indicators or your improvement plan is off target. Leading and lagging KPI's are powerful tools when used together.

Verification of controls

Let's consider dropped objects from elevated work in more depth. The concept is quite simple. The controls needed to prevent dropped objects are known. If we measure how well the controls are implemented **during** the work, then we should know the relative likelihood of a dropped object. Any dropped objects result from one of two broad causes, either the specified controls were not implemented effectively or a scenario was not anticipated. On the surface, this KPI provides insight only on the first cause, but inspectors can determine whether the scenario was identified or not through discussion with the work crew during the observation.

If safety inspectors, quality inspectors, or supervisors from another area survey several work parties every day and record whether effective controls are in place, you can develop simple charts of *Percent Implementation of Controls*. Examples of checkpoints may be:

- Are tool lanyards secured at both ends?
- Are gaps or holes in grating covered with tarps, boards, or other barriers?
- Is an effective exclusion zone established below the work?
- Are exclusion zones free from unauthorized or unnecessary personnel?
- Are scaffolds fitted with toe boards and free from excessive gaps?
- Are loose items present at the worksite?

Unlike open-ended worker engagement questions, these are simple Yes/No questions by design. A response of Not Applicable (NA) may also be appropriate. For example, the scaffolding question would be NA if the work was being done from a permanent platform. Examples of such checklists can be obtained via the download link in the Author's Note.

The results of these inspections will show whether controls are strong or need improvement. Of course, inspectors need to gather insights on **why** the deficiencies were present so improvement actions can be targeted to the cause. The inspections should not be a *tick the box* exercise from afar, but include a discussion with the work crew.

A simple scorecard showing your leading and lagging indicators can be used to share progress with the entire organization. As leaders, scrutinize it every week or month. As they say, *challenge the green and support the red*. If the scorecard consistently shows all controls are 99 percent effective, go out and check for yourself. I never believed such near-perfect performance. For the lower performing areas, ask why, show empathy, and provide the support needed to improve.

Audits

Audits offer broader insights into how to improve dropped object performance. They focus on the completeness and health of the systems in place. For example, perhaps the job-specific risk assessments are not done consistently across the facility, or maybe new employees are not trained for three months. Audit teams can be staffed with internal or external resources, or both. Auditors from outside of the facility provide a more objective perspective and often bring other relevant experience to share. Combining more frequent self-assessments with less frequent external audits is a good practice.

Improvement

Incident investigation and learning

Despite your best intentions and efforts, dropped objects will likely occur during the improvement journey, especially early on. Hopefully, they do not injure anyone. If you developed a good plan and provided enough resources, don't toss out the plan and start over after the first incident or two. Take the opportunity to learn from these incidents, no matter what the severity. Spend more effort on the serious incidents and the ones with high learning potential, perhaps a scenario not previously identified. These efforts may highlight gaps in the plan or opportunities to improve implementation.

The organization will scrutinize how leaders respond to incidents, so show supportive behaviors, such as:

- Emphasize that all dropped objects must be reported, even those posing low risk.
- Don't overreact; **learn** rather than **blame**.
- Recognize the efforts to report incidents.
- Track all incidents by type, craft, area, contractor, etc. to identify any trends.
- Use the Risk Calculator to classify the incidents; don't allow games to be played with the classification.
- Investigate all incidents to the depth dictated by the severity and learning potential.
- Identify the root causes.
- Hold formal Incident Review Panels on the higher priority incidents to obtain broader input.
- Integrate corrective actions into the improvement plan.
- Celebrate the lack of incidents and good leading indicators along the way.

You can also learn from the misfortune of others. Many industry and safety forums share lessons learned; some are listed in the Resources section at the back of the book. Resist the temptation to say "that could never happen here" or "we're better than that."

Learning involves more than sharing information at a safety meeting or in an incident alert. Corrective actions for root causes need to be embedded in the procedure, training, and other systems to be sustainable.

Management review

Most HSE Management Systems include a management review, typically an annual event where leaders review a wide variety of information relevant to overall HSE Performance. This concept can be adapted to create a Dropped Object Prevention Effectiveness Review. It may be a separate review or included as a focus area in the broader HSE review. There should be a plethora of information on your dropped object prevention program: incidents, leading indicators, audit results, safety observation trends, and DROPS Committee feedback. The DROPS focal point can compile and summarize the information for the leadership team to review. The complete picture may provide ideas for updating the improvement plan.

Summary

- Safety Management Systems (SMS) can bolster the effectiveness of dropped object controls at the front line.
- Sustainable improvement starts with leadership.
- Build a Dropped Object Prevention Culture by asking three fundamental questions: What could fall?, Where could it go?, and How can I control the risk?
- Use procedures, training, and contractor engagement to communicate expectations to the workforce.
- Provide the necessary materials, systems, and people to support the

front-line in their prevention efforts.

- Utilize leading indicators to monitor progress. Verification of the effectiveness of the controls at the front-line is a powerful tool.
- When incidents occur, learn; don't blame.

Personal reflection and learning

- How do you personally demonstrate your commitment to reducing harm from dropped objects?
- Which systems do you use at your site? How can they bolster the effectiveness of dropped object controls?
- Which system is not used, but is one you believe will help reduce dropped objects? Why do you think so?
- How well is the dropped object hazard covered in your procedure(s)? In your site orientation? In detailed training? Ask front-line workers what they have learned about dropped objects at the site.
- Visit a worksite with elevated work. Complete a dropped object inspection checklist with the work crew. Which controls were present but ineffective? Which controls should be added?
- What data (leading and lagging) do you see related to dropped object performance? What does it tell you? What are you missing?
- Plan and execute an awareness-building campaign or activity
- Organize a hazard hunt focused on dropped objects. What does the team find? How do they react? How do you react?
- Hold a Dropped Object Prevention Effectiveness Review.

9

Industry-Specific Challenges

Some industries have unique challenges or one type of dropped object dominating their incidents. If you are in one of these industries, the summary below may help focus your improvement efforts. If not, it may help you discover best practices for challenges also present in your business.

Even within an industry, the risk profile varies from site to site and time to time. From site to site, equipment, production methods, contractors, safety culture, languages, and regulations may vary. Over time, a site may face challenges due to startup, labor relations, change in ownership, expansion, and concentrated maintenance periods. These factors can all affect dropped object exposure and should be considered when developing an improvement plan.

Oil and gas drilling

This industry has high exposure for reasons identified below. They have been at the forefront of developing dropped object prevention practices. To expedite improvement through open sharing of best practices, they founded the DROPS organization (Dropped Object Prevention Scheme). Other industries have benefited greatly from this sharing of tools and lessons learned.

Elevation differences are significant since drilling derricks handle long pipes (or tubulars) used in well construction. The derricks support heavy, overhead moving equipment consisting of many components that can become loose. The equipment can also cause vibration for the rest of the structure. These factors led to the rigorous Reliable Securing and inspection programs described in Chapter 6.

Workers are often present near the derrick to control and adjust equipment and well components. Rigs typically have rigorous zone management programs with different degrees of exclusion zones, DROPS shacks for overhead protection, and detailed procedures describing where people should be located during different operations. The industry has automated many operations to reduce the exposure.

Drilling uses a lot of tubulars (drill pipe, casing) and consumables (mud, chemicals), so material handling is another significant contributor to dropped objects. Tubulars are removed from a truck or supply boat and placed in a staging area. They are moved to the derrick when needed for drilling, completing, or servicing the well. Some tubulars are left in the ground, but others must be handled in the reverse order as the operations are completed.

Many drilling rigs use dedicated *Work At Height toolkits*. Overhead work is not extensive during drilling operations, so this control is feasible. Major construction and repair projects are usually done between drilling operations.

Companies move rigs from location to location, whether it be breaking down and re-erecting a land-based unit or moving a drill ship from one location to another offshore. This leads to additional overhead operations and connections, especially for land-based units.

Offshore facilities

Most of these facilities are used for oil and gas drilling and production. They operate in challenging environments with corrosive sea spray, high winds, pounding waves, and strong currents. If the structure is floating, the wind, waves, and currents lead to structural motions that can cause equipment to work loose. Reliable Securing and inspections/maintenance can reduce this risk. Effective implementation of this control can be challenging because of limited staffing, which is often minimized to reduce the overall facility risk and cost.

Elevations are moderate, but congestion of equipment is high due to the expense of creating the large structures needed to support the equipment. Exclusion zones should be large enough to account for multiple deflections but not block emergency egress routes.

Lifting and hoisting activity is high since most supplies arrive on supply vessels. Loads are generally well-secured, but mistakes occur. Laydown areas are limited, so careful planning is needed to reduce multiple handling of equipment and containers.

Wind power generation

Dizzying heights dominate the dropped object risk profile. The typical height of an onshore wind turbine is just under 300 feet. At this height, the risk calculator shows the weight of potentially fatal dropped objects approaching **zero**, so any falling object is dangerous. Rigorous use of tool lanyards and tool and material accounting are essential.

Most of the facilities are in remote areas and well-spaced, so fewer people are present. However, it may be hard to control access in these wide-open spaces.

The number of wind turbines offshore is growing. Even fewer people are present, but they face similar environmental challenges as offshore oil and gas facilities.

The Global Offshore Wind Health and Safety Organisation (G+) has customized some of the best practices developed by DROPS to focus on their unique exposures. This is a great example of cross-industry learning. (http://www.gplusoffshorewind.com/work-programme/workstreams/guid elines)

Large manufacturing facilities

All three types of dropped objects are prevalent in large manufacturing facilities, whether producing gasoline, electricity, lumber, or automobiles. The biggest challenge is managing all three well. The key to success is determining which aspect presents the most risk so improvement efforts can be focused.

Periods of intense maintenance work (sometimes called *turnarounds*) and construction work can significantly increase the risk of dropped objects from elevated work. Tools and materials left behind from such efforts can linger for years before falling.

Material movements are extensive, using most modes of transportation and lifting.

Many of these facilities are quite old. Falling overhead equipment can become a larger contributor to dropped objects as the facility ages if rigorous inspection and maintenance programs are not in place. This control can be challenging to implement at large facilities. The overhead equipment is vast, and many inspections are required to manage a variety of risks. At facilities producing solids or durable goods, overhead moving equipment further increases the risk.

Layouts are a mix of congested production areas and less congested support areas, such as tank farms, pipelines, laydown areas. Elevation differences can be large in the production areas, especially around the perimeters.

Warehousing

Material movement dominates the risk at these facilities. Controls include well-designed storage racks, proper containers, load securement, equipment operator competency, and organization/planning.

In most cases, material arrives and departs by truck. Many people underestimate the risk of dropped objects from trucks due to the low height. Prior engagement with suppliers to specify acceptable containers and rigging can reduce deliveries of unsafe loads that are turned away, or worse yet, unloaded anyway.

Private construction

Extensive elevated work and material movements lead to a very high risk of dropped objects at large construction sites. Private construction sites are typically able to control access to the site, whereas public construction is typically done in areas which the public may access, for example, construction in downtown areas and road construction.

Unlike for routine operations, work areas are always moving. An area with low risk last week may be high risk this week. Most lifting equipment is mobile. Scaffolding work is significant.

With high exposure in elevated work, verification of effective implementation of controls can be an effective tool to reduce risk. A few shortcuts here and there add up to enormous exposure across a large site.

Exclusion zones can be difficult to manage. Barricades seem to be everywhere on busy, congested sites and can prevent people from getting to or from their work location. This is a real problem in an emergency, but it also frustrates workers and encourages them to pass through barricades.

Contractors are mobilized and demobilized as different trades are needed, so dropped object prevention knowledge must be instilled quickly. Engagement with contractors prior to arrival can help prepare them to manage the risk from the first day. Ideally, their own procedures are consistent with the site's procedure, and their people are trained before arrival. In other cases, they may use the site's procedure and training program. Increased supervision and coaching by safety professionals during the first few weeks can also accelerate onboarding.

Material handling is significant. Many types of third-party heavy equipment are used. Site staff should verify that it is fit for purpose and inspected and maintained per manufacturer's recommendations. Competence of the equipment operators and rigging personnel is also an important control.

Public construction

One big difference between private and public construction is site access. Members of the public will typically be less aware of the risk of dropped objects than those on a private site. This leads to extreme measures to *Obstruct* (shrink-wrapped scaffolds) and *Prevent* (walkways covered with hard tops).

Summary

- Industries face different challenges in preventing dropped objects. Learn from those with challenges similar to yours.

Personal reflection and learning

- What dropped object scenarios are unique to your industry? Which controls are needed to prevent them? Are new ones needed?
- What other industry has similar challenges to yours? How can you learn from that industry?

10

Where Do I Start?

5 steps to reducing dropped objects

Congratulations! You should now understand:

- The scope of the dropped objects problem.
- How to estimate the risk of potential dropped objects.
- The different types of dropped objects.
- The controls used to prevent dropped objects.
- How to use a Safety Management System to bolster the controls.

"This is great, but what do I do now? You shared so much material. Do I need to implement everything? Where do I start?" These are all good questions. The key to success is **focus**.

The improvement journey can be broken into **5 Steps**. These 5 steps will work no matter how large your problem is; it will just take longer for broader scopes. The steps are easy to understand, but require hard work to implement. But as mentioned before, so do most worthwhile things in life.

Readers of this book will be in different types of roles. Leaders of large companies are learning how to drive safety improvement in their entire organization. Leaders of teams are learning how to keep their own team members safe. Safety managers and staff are preparing to support front-line workers and their leaders. They may also be garnering the support of a site leader who is not on board yet. Other individuals are showing initiative to keep themselves and their coworkers safe after hearing all the talk about dropped objects.

I will first focus on the target leaders in an organization identified in the Introduction, i.e., those two to four layers above front-line workers. Readers in other roles should see where they fit, but recommended actions for other roles will follow.

The **5 Steps** are:

1. Identify your **focus** area.
2. Document and **communicate** mandatory requirements.
3. **Resource** the plan.
4. **Lead** through coaching and verification.
5. **Learn** and adjust.

1. Identify your focus area

As McChesney, Covey, and Huling state in their book, *The 4 Disciplines of Execution*, "the more you try to do, the less you actually accomplish." If too many changes are attempted at once, none may be successful. First, check the amount of change in progress, not just in safety, but across the board. Leaders and staff have limited capacity for significant change, so you may need to limit yourself to two or three at a time. Second, are dropped objects one of the top safety risks at the site? If so, you can select one type of dropped object to focus on first, probably the one posing the most risk to workers. Once you make significant and sustainable improvement there, you can focus on other types of dropped objects, another safety hazard, or the next business priority on your list. If the best opportunity for safety improvement lies elsewhere, then you may want to focus there first. These five steps can be applied to any safety improvement initiative.

Notice the discussion on focus is about changes. Workers encounter dozens of safety hazards every day. They still need to manage those risks, or else they may be hurt in other ways. Just like you must still perform the daily tasks needed to run the entire business. The focus on dropped objects, or whatever area you select, is to use that little leftover capacity to make improvement in a specific area.

"How do I select my focus area?" Create a diverse team to help frame the problem for your organization. Team members from Safety, Maintenance, Inspection, Operations, Logistics, and Contractor Management add valuable perspectives. Ask them to learn more about the risk using this book, referenced materials, training, and other resources.

The team should then analyze the site's performance data (hindsight) and exposure (foresight). What types of dropped objects are most common, especially the more serious ones? What does the exposure look like over the next year? Is a large turnaround scheduled soon? How much construction is on the horizon? Is the facility showing its age? What are inspectors observing during their routine equipment inspections? The answers to these questions should make the focus area clear.

The focus area is likely to be dropped objects from **elevated work** if:

- Most dropped object incidents are related to elevated work.
- A large maintenance campaign is coming soon.
- Project or construction work is high or increasing.
- The facility is new or has been well-maintained.

The focus area is likely to be dropped objects from **elevated equipment** if:

- Most dropped object incidents are related to elevated equipment.
- Little maintenance or construction work is planned.
- The facility is old or has not been well-maintained.
- Reports of loose equipment are increasing.

The focus area is likely to be dropped objects from **material movement** if:

- Most dropped object incidents are related to material movement.
- Lifting and material handling programs are not robust.
- A lot of maintenance or construction is coming soon.
- The facility is new or has been well-maintained.

Don't forget to consider the systems. If you have a good suite of controls in all three areas, but implementation is lagging, you can focus on a system or two instead.

If all three types appear to be equally important, just select one to begin the journey, perhaps the one that will be easier to improve so you achieve a quick win to help propel improvement in the others.

Consider bringing in a consultant or someone from another part of your organization to advise. They will bring insights and lessons learned that you are too close to see.

Finally, once the focus area is identified, review it with your Leadership Team to gain alignment. You will be asking for significant changes in the way their groups work, so they must be on board.

2. Document and communicate mandatory requirements

The primary tools to communicate expectations on dropped object prevention are the procedure, training, and contractor engagement. For your focus area, identify the controls front-line workers are expected to use and the systems needed to support the controls. Document these in a procedure (or plan), which will be the basis of the training. Share this with contractors to solicit their input and support.

Note the word, *mandatory,* in the subtitle. The procedure and training should emphasize that dropped object controls are required, not suggested, recommended, or nice to do. Optionality is a big reason so many things still fall. Sure, workers need flexibility in how they apply controls in specific work scenarios, but the basics must be in place. The inability to meet these requirements in unusual situations should be approved through a deviation process documented in the procedure.

A comprehensive dropped object procedure should be completed eventually. If you can't complete it quickly, at least finish the parts applicable to your focus area.

Timing is important. Before you start training everyone, make sure the resources are in place to support it. If not, you will lose their interest and your credibility.

3. Resource the Plan

For elevated work, provide the resources and materials needed for the controls: tool lanyards, rigging equipment, tarps, boards, netting, barricade tape and signs–whatever the procedure requires. Make it easy for front-line workers by storing materials close to the work and ensuring the materials are always available. But additional resources will be needed depending upon the focus area. They include:

- Risk assessment tool
- Awareness campaign materials
- DROPS focal point(s)
- Equipment securing guidelines

- Inspection protocols
- Improved rigging components
- Load securement supplies
- Inspection forms for verification
- Inspection staff

4. Lead through coaching and verification

This is number four on the list, but number **one** in importance. Leadership is embedded in all five steps; it just happens to be in the title of this one. Initiating the change may seem difficult, but is often the easiest part. Nurturing the change and maintaining the focus is the hard part for many leaders. But if you don't persist, the organization may conclude it is just another brilliant idea from the ivory tower that will pass.

Good safety leaders make this look easy. Let's go back to my story about responding to those three serious dropped objects in one week and the change to a formal safety verification process.

> The Project Manager kept the focus on dropped objects visible for three years until the very last day of the project. In addition, the effect rubbed off on the operations leaders and staff to which we delivered the asset, and they had few dropped objects in the following years.

> Her actions were simple when considered individually. The difficult part was the consistent emphasis over the rest of the project, while the team faced many unique challenges, even crises. She insisted on monthly meetings with her leadership team. The work sites held their review meetings weekly. Those meetings were closer to the work and more important, but she wanted her leadership team to demonstrate commitment and help the sites learn from

each other. As HSE Manager, I gathered the verification data from the work sites, and we reviewed it during the meetings.

She asked probing questions about any trends. "I see Site A has improved their use of tool lanyards. How did they do that? Now the effectiveness of their exclusion zones is deteriorating. What are they doing about that? Is there anything we can do to help?" If there were no new trends, "What is the next challenge? What will we do about it?"

She focused on dropped objects more than any other single topic while visiting project sites. She asked about the challenges the site faced and what was working well, then went to the workplace to see and hear for herself.

She challenged her leadership team to replicate effective work practices or controls at all the project sites. She influenced other projects and other operating units at the plant to implement our learnings. She recognized the team for their individual and collective success in front of each other and senior leadership of the company.

She led like this for **three years**. I have tried to implement changes like this in other projects and operations, enough to realize that without this commitment and support, the change would have faded away. The dropped objects would have continued, and someone would be hurt.

5. *Learn and adjust*

Dropped objects may continue as you try to improve, especially early on. Unless something relevant has changed, stick with the overall plan, learn from those incidents, and adjust as needed. Because of the intense focus on dropped objects, everyone will watch your response to incidents. After all, they know it is important to you and don't want to disappoint you. You can respond in one of two ways: find someone to blame or **learn** from the incidents. Blame will erode trust and drive reporting underground, including the learnings that come with the incidents. Learning will reinforce your genuine commitment to create a safer workplace and increase your ability to influence others.

When can you move on to the next problem?

Dropped objects will not be your only safety challenge. And safety will not be your only leadership challenge. Your focus will need to shift over time to deliver other commitments on safety, production, schedule, costs, quality, people development, etc.

When can you shift your focus to another priority? There is no clear answer, but it may be longer than you think. First, have the incidents declined significantly? That is the outcome you are committed to. If not, either the controls have not been embedded, or some scenarios were not foreseen. There is still work to do on dropped objects.

After the incidents decline, even to zero for a short period, it still may not be time to switch your focus. You may have been lucky for a while. What do the leading indicators show? Are the controls getting better? Both leading and lagging indicators should show significant improvement before you move on.

A good way to test your readiness to shift focus is to back off on your inquiries and meetings and observe what happens. If you stop asking about dropped objects during site visits, do the site leaders still share their successes and challenges anyway? Are they still tracking leading indicators? Are the workers still using the expected controls? If you postpone the performance review meetings, does the data keep coming in? Are any negative trends addressed without your inquiry? Do team members still come to you with good practices to replicate?

If performance has improved and the answer to most of these questions is **yes**, then you should be able to focus on another type of dropped object or another safety or business challenge. If **not**, then you may need to maintain the focus and perhaps supplement it with coaching team members who are not fully embracing the change.

Don't assume building robust controls for one type of dropped object will prevent all dropped objects. The controls are different and are implemented by different people. If the controls for other types of dropped objects need improvement, give them some focus as well once the organization has some capacity. Potentially serious dropped objects may still be likely until those improvements are made.

In conclusion, you never really *drop it*. The risk assessments continue in the field, the team leaders still use leading indicator data, and good practices are shared naturally. But you and your team now have the capacity to take on a new challenge. Those new tools and controls have become the normal way of working. You built a dropped object prevention culture. Since it is culture, it should take a long time to degrade if left unattended for a while. You can always maintain a leading indicator or two to give an early warning if backsliding occurs.

Everyone can help

I hope every one of my readers found useful information here that will help them prevent harm from dropped objects. There is plenty of work to go around. Since the last section addressed the 5 high-level steps that target leaders in an organization should take, here are more specific actions for them and people in other roles. I haven't covered every imaginable role below, but the lists are long enough that I am confident you can find a few actions for your specific role.

Leader of a very large site or organization

- Determine the priority of dropped object prevention within the long list of improvement opportunities for the organization.
- Provide resources for the improvement program.
- Review progress and performance regularly in leadership team meetings.
- Recognize excellent dropped object prevention behaviors.
- Chair incident review panels to ensure root causes and learning opportunities are identified.
- Visit the worksite and test the controls.
- Participate in dropped object awareness events.
- Influence business partners, including contractors, to commit to and participate in the improvement program.
- Sponsor a DROPS Committee to receive feedback on dropped object prevention in practice. Follow-up on recommendations or requests for assistance.

Safety Manager/Staff

- Gather and analyze leading and lagging indicators to monitor progress.
- If the line leaders are not taking action, make a case for action.
- Identify a focus area to start the improvement journey.
- Provide resources for the DROPS focal point and trainers; provide safety

inspectors to coach the front line and collect leading indicator data.

- Draft a dropped object prevention procedure.
- Develop training materials.
- Develop and deliver awareness programs.
- Organize the annual Dropped Object Prevention Effectiveness Review.
- Organize a peer review or audit.
- Investigate root causes of dropped objects; adjust the controls and systems.
- Visit the worksite and test the controls yourself.
- Connect with industry organizations and other safety managers to share challenges and successes.

Safety Inspector

- Continuously learn about the risk and controls.
- Observe and coach the front line on the risk and controls; intervene when necessary.
- Collect leading indicator data; ask questions to understand the causes.
- Plan and execute awareness events.
- Participate in pre-job dropped object risk assessments and toolbox talks.
- Collect ideas for improvement in controls from the front line; share with other teams and leadership.
- Contribute to development of the dropped object prevention procedure.
- Educate contractor leaders and safety inspectors on the risk and the controls.

Front-Line Supervisor

- Continuously learn about the risk and controls; understand the requirements in the procedure.
- Make sure your crew is trained on dropped object prevention.
- Lead specific discussions about dropped objects in toolbox talks; ask open-ended questions.

- Ensure pre-job risk assessments identify potential dropped objects and robust controls.
- Visit the job site regularly to verify the identified controls are in place and effective.
- Stop work if your team members cannot implement effective controls; help workers resolve dilemmas.
- Recognize those who go above and beyond their responsibilities.
- Report and collect initial information on dropped object incidents, even if low risk.
- Collect ideas for improvement in controls; share with other teams and leadership.

Front-Line Worker

- Learn about the risk and controls - actively participate in training.
- Contribute to discussions about dropped objects in the toolbox talks.
- Complete a pre-job risk assessment to identify the correct controls.
- Share ideas for improvement in controls with your supervisor or safety inspector.
- Stop work if you cannot implement effective controls; contact your supervisor for support.
- Take only necessary tools and materials to height; complete testing or pre-assembly at ground level.
- Obtain supplies for controls before going to height.
- Ensure no one is working above or below you.
- Report any dropped objects, even if low risk.
- Do not enter exclusion zones unless you have permission, and it is safe to do so.

Operations/Production Manager

- Visit the worksite and test the controls.
- Establish a budget for supplies for the controls.

- Organize and participate in DROPS hazard hunts.
- Provide personnel for the DROPS Committee.
- Encourage operators to look for and report potential dropped objects from elevated equipment.
- Have your team set the example of a dropped object prevention culture by securing personal accessories.
- Recognize excellent dropped object prevention behaviors.
- Sponsor incident investigations; ensure they identify root causes.

Maintenance/Construction Managers

- Visit the worksite and test the controls.
- Establish a budget for supplies for the controls.
- Organize and participate in DROPS hazard hunts.
- Ensure repairs aimed at potential dropped objects are in the maintenance management system and properly risk-ranked.
- Track potential dropped object repairs in the maintenance backlog.
- Engage with contractor leaders to ensure their understanding of dropped object prevention requirements.
- Encourage/require contractors to write their own dropped object prevention procedure specific to their scopes of work.
- Provide personnel for the DROPS Committee; encourage contractors to participate.
- Work with scaffolders to ensure they are using best practices.
- Make equipment available to reduce manual handling of materials.
- Establish a process for choosing the best means of access for elevated work.
- Identify opportunities to reduce elevated work.

Inspection Manager

- Establish systematic and focused inspection schedules and protocols for preventing dropped objects from elevated equipment.

- Conduct inspections on time.
- Verify the quality of inspections.
- Identify what good looks like for reliably securing common types of equipment.
- Visit the worksite and observe the condition of elevated equipment and securing methods used.
- Provide personnel for the DROPS Committee.

Logistics/Lifting Supervisors

- Ensure equipment is inspected and maintained per manufacturer's recommendations, even when demand for the equipment is high.
- Visit the worksite and observe the condition of equipment and the lifting/transport practices.
- Ensure procedures and training include regulatory requirements and best practices.
- Make equipment available to reduce manual handling of materials.
- Provide personnel for the DROPS Committee.

Project Engineers

- Systematically evaluate options to reduce elevated work in construction, operations, and maintenance.
- Make equipment available to reduce manual handling of materials.
- Ensure contractors arrive at the site prepared to prevent dropped objects.
- Ensure the dropped object prevention procedure covers the scope of your project; if not, consider a project-specific plan or procedure.

Everyone

- Learn about the risk and the controls.
- Go out to the worksite and discuss the risk with the front-line.
- Recognize those taking proactive steps to prevent dropped objects.

- Think about how else you could support the effort.

Summary

- Focus on a small number of improvement initiatives to improve the probability of success.
- These five steps can help you improve: (1) Identify your **focus** area, (2) Document and **communicate** mandatory requirements, (3) **Resource** the plan, (4) **Lead** through coaching and verification, and (5) **Learn** and adjust.
- Committed, involved, and consistent leadership is essential, but everyone can and should contribute.
- Make sure the improvements are embedded and sustainable before moving on to the next focus area.

Personal reflection and learning

- What are your top three safety improvement priorities? What other improvement initiatives are in progress?
- What type of dropped object is your highest priority for improvement? Why?
- What two or three actions do you think would reduce dropped objects the most at your site?
- Who do you need to influence to gain support for changes? How will you do so?
- What can you do personally to support the improvement journey?

11

Conclusion

"Wow, I never imagined there was so much to know about dropped objects. It will take me forever to do all of this. Maybe I should work on something else first?" To that I say, just get started. Too much is at stake. What else are you doing that is more important?

You can make a difference, whether you are the top leader at a facility, the person building the scaffold, or the project engineer building a new tank. If you are the one at the top, you have great responsibility but also significant influence and control of resources. If you are the scaffolder, you can enable your colleagues and the people who will be working from your scaffolds to go home unharmed. If you are the project engineer, you not only have the responsibility to prevent dropped objects during construction, but also the opportunity to reduce exposure to dropped objects for operations in the future.

If you feel overwhelmed, think about this:

- If your exposure is high, someone will probably be hurt sooner or later. It could be tomorrow. You will be accountable for the consequences and will live with them forever. Don't wait until it is too late.
- Others have reduced this risk significantly. If they can do it, you can too,

probably even faster since you know how they were successful.

- "You can't finish what you don't start...," said Gary Ryan Blair. So, get started. You can focus on one area, get it right, and then move on to the next.
- Leadership drives change. If you don't lead it, who will?

Let's revisit those excuses in the Introduction.

- *"Our equipment is not so high. Dropped objects are not a problem here."* The risk calculator helps everyone understand the risk. Heavy things falling short distances can be deadly. The risk is everywhere.
- *"That could never happen! Nothing could fall from here."* I shared a few examples that should convince you this mindset is dangerous. Ask around, and you will find more. If you anticipate the scenario, you can prevent it. If you ignore it, someone will get hurt.
- *"But we are not working at height. We are on a permanent platform."* Dropped object prevention is not the same as personal fall prevention. Smaller objects fall in ways that people do not, for example, through small gaps in work surfaces.
- *"All of our people use tool lanyards. There are plenty in the tool room."* Just because you bought them, doesn't mean they are being used, or used correctly. Plus, tool lanyards are just one of many layers of protection needed.
- *"The area was barricaded. No one could have been hurt."* Barricades are an administrative control, one of the least effective in the hierarchy of controls. But they are important and can be improved. People make mistakes, so fail safely by using multiple layers of protection.
- *"My inspectors will find any potential DROPS while performing their structural inspections. Why change anything?"* Inspectors focus on specific problems by design. They may find unrelated problems, including loose items at height, but you cannot count on it.

Let's also revisit those reasons why preventing dropped objects is difficult.

It is still difficult, but you now understand why and have tools to face the challenge head-on.

- *It's not just one problem*: You know the three types of dropped objects and controls for each. I have also shown the importance of focus and how to select a focus area.
- *There is no rulebook yet*: You have seen the controls and systems to include in your own rulebook. I have also provided links to recommended practices from industry organizations in the Resources section.
- *The risk is not intuitive*: You learned about the Dropped Object Risk Calculator, a very simple tool to help everyone in your organization understand the risk.
- *Everyone has to get it*: The risk calculator helps here as well. I gave examples of other techniques to build an appreciation of the risk. I also described how everyone in your organization can help with the improvement journey.
- *Everything can fall*: You now know the three types of dropped objects and many of the mechanisms of failure. I have also provided numerous examples which should spur you to think of others within your own business.
- *Gravity never sleeps*: Sorry, neither you nor I can make gravity disappear. But I have covered the need for multiple layers of protection so that you can fail safely in most scenarios.
- *Safety is always hard*: I used the word, *always*, for a reason. That's why we all need to keep learning. Many of the SMS elements covered in Chapter 8 will help reduce the risk of many hazards in your business. Good leaders solve hard problems.

Don't think you need to do everything. The practices presented here have been successful at multiple sites. However, no one site is doing all of them. You will find some more effective than others for your culture and your risk profile. For now, commit to at least two actions: (1) getting started with the basics and (2) improving continuously. You won't regret it.

Do your job. Get started. Save a Life!

12

Resources

Dropped Object Prevention Scheme (DROPS)

DROPS (in this context) is short for Dropped Object Prevention Scheme. This organization was formed by the oil and gas drilling industry in 1998. This industry has a challenging risk profile and collaborates to reduce dropped objects. They initially focused on the key issues in their industry, but over time, they have broadened their scope. DROPS has developed many tools and best practices that are helpful to everyone and free to use and adapt, including the dropped objects risk calculator, Reliable Securing, and a dropped object survey template. They also share a Recommended Practice, incident reports, and other materials, and offer training to increase awareness. An excellent resource! (http://www.dropsonline.org/)

Global Offshore Wind Health and Safety Organisation (G+)

G+, the Global Offshore Wind Health and Safety Organisation, in partnership with the Energy Institute, has adapted the DROPS Reliable Securing document to the growing offshore wind energy industry (Reliable Securing Booklet for Offshore Wind).(http://www.gplusoffshorewind.com/work-pr ogramme/workstreams/guidelines).

American Clean Power

American Clean Power, formerly known as the American Wind Energy Association (AWEA), created a Stop the Drop campaign in 2015 containing best practices, posters, and videos focused on dropped object prevention, mostly related to elevated work.(http://cleanpower.org/environmental-heal th-and-safety/safety-campaigns/prevention-of-dropped-objects/)

International Oil and Gas Producers Association (IOGP)

IOGP publishes many recommended practices to promote consistent use of safe work practices within their membership companies and their business partners. Recommended Practice 577: Fabrication Site Construction Safety Recommended Practices specifically addresses dropped objects and many other construction safety risks, such as falls from height, hot work, confined space entry, and lifting and hoisting. It recommends that companies develop their own specific dropped object prevention plans. As you might expect, it focuses mostly on dropped objects from elevated work. (http://www.iogp.o rg/oil-and-gas-safety/construction/)

Energy Institute

Energy Institute, the chartered professional membership body of the energy industries, provides the following resources on dropped objects free-of-charge:

- The Stop Drops reflective learning toolkit includes a series of videos and a facilitator guide covering the different types of dropped objects presented in this book. (http://heartsandminds.energyinst.org/toolkit/ reflective-lfi#Vid6)
- Toolbox contains lessons learned and other safety information in a simple format useful in preparing for toolbox talks. The materials cover a wide range of hazards, including dropped objects. (http://toolbox.ene

rgyinst.org/home)

Energy Safety Canada

This energy industry association provides tools and learning materials on dropped objects, including a short but excellent microlearning video illustrating the types of dropped objects. They have also developed a draft tool to help estimate potential deflection distances of dropped objects. (http://www.energysafetycanada.com/Standards/Programs/Dropped-Obje cts)

Vendors for dropped object prevention supplies

Many vendors supply companies with fit-for-purpose, rated, and tested equipment to prevent dropped objects. Their products include but are not limited to tool lanyards, spill-proof bags, guardrail covers, and exclusion zone markers. They can provide practical advice and training on using their products and may be able to develop a solution for unique materials that are difficult to secure at your site.

American National Standard For Dropped Object Prevention Solutions

The ANSI/ISEA 121-2018 standard "establishes minimum design, performance, testing and labeling requirements for solutions that reduce dropped objects incidents in industrial and occupational settings." (https://webstore. ansi.org/Standards/ISEA/ANSIISEA1212018?gclid=CjwKCAiA1aiMBhAU EiwACw25MZuPGQEiROORwFAAvk40UAJ2PmJPXwfrULeEVvh2LGacr WRme1yNjRoCWWUQAvD_BwE)

Melon Drop Demonstration

Black and Veatch share an example of the *melon drop* demonstration in this video, the Dropped Object Experiment. Objects of various weights and shapes are dropped down a tube onto a watermelon with and without the protection of a hard hat. This type of demonstration creates a lasting impression on the workforce. (http://www.youtube.com/watch?v=6wQKU DX7D94&t=2s)

National Access and Scaffolding Confederation (NASC)

NASC is a trade body for the industry based in the UK. It has published several guides to improve safe work practices for its members and the industry. Examples related to Dropped Objects include: (http://nasc.org.uk/)

SG9:21 *Use, Inspection & Maintenance of Lifting Equipment and Accessories for Lifting in Scaffolding*

SG6:15 *Manual Handling in the Scaffolding Industry*

ISO 45001

ISO 45001:2018 - Occupational Health and Safety Management Systems contains requirements of an effective (Health and) Safety Management System, along with guidance for application in businesses. This reference was used to organize Chapter 8, which described systems and tools that can increase the effectiveness of controls at the front-line. A guidebook for application in small businesses is also available.(http://www.iso.org/iso-450 01-occupational-health-and-safety.html)

Author's Note

I hope you found this book enjoyable, motivating, and useful in planning your next steps to reduce dropped objects and save lives. Please **leave a review** where you bought this book to help others do the same. One of the biggest challenges for new authors is discoverability, but your reviews and referrals will help. Thank you in advance.

You can download the free tools referenced earlier here. (https://drive.goog le.com/uc?export=download&id=1e5140aCJyl_2xkF6EW9BtO0ZQcPUTd 1R).

For updates on new resources and tools that support this book and upcoming books and events, please sign up for my email list on my website here. (https://strivingforsafety.mailerpage.com/).

Please encourage others in your network to read the book and sign up for the email list. This leads to more people working on the problem and more people to learn from later.

My LinkedIn profile can be found here. (https://www.linkedin.com/in/arn old-marsden-14170180)

You can email me at arnold@strivingforsafety.com.

Best wishes on your journey!

Notes

WHY IS IT SO HARD?

1   Survey of Occupational Injuries and Illnesses Data, Bureau of Labor Statistics, November, 2021. https://www.bls.gov/iif/oshwc/osh/case/cd_r31_2020.htm

2   Census of Fatal Occupational Injuries, U.S. Bureau of Labor Statistics, December, 2020. https://www.bls.gov/iif/oshcfoi1.htm#2019.

HOW BAD COULD IT BE?

3   DROPS Calculator, Dropped Objects Prevention Scheme, October, 2021. https://www.dropsonline.org/resources-and-guidance/drops-calculator/.

THREE TYPES OF DROPPED OBJECTS

4   DROPS Recommended Practice, Dropped Objects Prevention Scheme, Revision 2, March 2020. https://www.dropsonline.org/resources-and-guidance/dropped-object-prevention-scheme-recommended-practice-rev-02/.

DROPPED OBJECTS FROM ELEVATED WORK

5   Dropped Objects Exclusion Zone Tool, Version 2.2, Energy Safety Canada, 2020. https://www.energysafetycanada.com/Standards/Programs/Dropped-Objects

DROPPED OBJECTS FROM ELEVATED EQUIPMENT

6   DROPS Reliable Securing, Revision 4.0, DROPS, July 2017. https://www.dropsonline.org/resources-and-guidance/drops-reliable-securing-booklet-rev-04/

7   DROPS Reliable Securing, Revision 4.0, DROPS, July 2017. https://www.dropsonline.org/resources-and-guidance/drops-reliable-securing-booklet-rev-04/

8   DROPS Reliable Securing, Revision 4.0, DROPS, July 2017. https://www.dropsonline.org/resources-and-guidance/drops-reliable-securing-booklet-rev-04/

9   DROPS Reliable Securing, Revision 4.0, DROPS, July 2017. https://www.dropsonline.org/resources-and-guidance/drops-reliable-securing-booklet-rev-04/

10   DROPS Common Guidelines for Dropped Object Surveys and Inspections, DROPS, October, 2011. https://www.dropsonline.org/resources-and-guidance/survey-and-inspection-english/

DROPPED OBJECTS FROM MATERIAL MOVEMENT

11   Use, Inspection & Maintenance of Lifting Equipment and Accessories for Lifting in Scaffolding, SG9:21, National Access and Scaffolding Confederation, May, 2021. https://nasc.org.uk/product/sg915-use-inspection-and-maintenance-of-lifting-equipment-and-accessories-for-lifting-in-scaffolding/

PUT YOUR SAFETY MANAGEMENT SYSTEM TO WORK

12   Occupational Health and Safety Management Systems, ISO 45001:2018, International Organization for Standardization, March, 2018. https://www.iso.org/iso-45001-occupational-health-and-safety.html

About the Author

Arnold was a Health, Safety, and Environment (HSE) professional for an international oil and gas company for over thirty-five years. His last position was Global Dropped Objects Focal Point, a role created to reduce the largest contributor to the company's serious near misses. He helped businesses around the world share and learn best practices and drove continuous improvement. High risk dropped objects fell by more than 30 percent in two years. Prior to that, he was the HSE Manager on several multi-billion dollar global projects. In that role, he worked with dozens of business partners around the world to prevent harm to workers in projects filled with serious hazards. He had the opportunity to observe safety leadership skills of a diverse group of leaders and coached them and their teams on how to become more effective safety leaders.

He enjoys sharing his knowledge with others, continuing to learn, and helping others improve. He has retired from the corporate world, but continues to help others through networking, coaching, and writing books like this one.

Arnold has also written a book on safety leadership, entitled Safety First! Really? How to be a Credible Safety Leader.

He also writes outdoor adventure stories with his Bucket List Hike series. The first in the series, Muir Trail Magic, is set on the 211-mile John Muir Trail in the Sierra Nevada in California.

You can discover his other books and follow his author journey here (https://s

trivingforsafety.mailerpage.com/).